Einstein Light Time Relativity

Assumptions of Science

by

Walter R. Dolen

Einstein Light Time Relativity:

Assumptions of Science

by

Walter R. Dolen

PARISBURG PUBLISHING
parisburg.com

First published edition as *Einstein: Light, Time and Relativity* (2015)
A few parts of this book were originally included in the author's non-published work, *The Totality of Motion's Relativity* (1971-72)

Graphics:

Front cover photo (c. 1905) in the public domain, photographer
 unknown;

Back cover photo is in the public domain and was the official 1921 Nobel
 Prize photo;

Fig. 1, unknown graphic artist found in Albert Einstein's *Relativity: The
 Special and the General Theory*, in Chap IX (Section 9), 1920
 English translation;

Fig. 6 are by Stan Freitas, found in the unpublished work, *The Totality of
 Motion's Relativity*, 1971-72, by Walter R. Dolen;

Fig. 3, 8 & 9 and graphic by Walter R. Dolen;

Fig. 2, 5, 7 & 10 are in the public domain under Wikipedia Common.

Fig. 4 Public domain, webbtelescope.org

ISBN: 978-1-61918-**069-7**
Hardback

ISBN: 978-1-61918-**068-0**
Paperback

Second Edition
First Printing
September 2025
(Digital version to follow)

This book was first published in the 100th year since Einstein published his "The Field Equations of Gravitation," for the General Theory of Relativity and 110th year since his 1905 relativity paper. This book 2025 is a second edition and is a constructive critique of Einstein's relativity theories and those theories that rely on it for their foundation.

PARISBURG PUBLISHING
parisburg.com
Pennsylvania, USA

Dedicated to:

This book is dedicated to all those who question conventional thinking and imagine outside the prevailing mindsets. This includes Albert Einstein, James Clerk Maxwell, Henri Poincaré, Nicolaus Copernicus, Johannes Kepler, Galileo Galilei, Isaac Newton, Hendrik Antoon Lorentz, Herbert Dingle, Paul Dirac, Halton C. Arp and many others. All of them questioned the theories of their day and presented alternate views. Scientific theories are only temporary vehicles used to drive closer to the truth of all things. True science leaves the disproven behind, and moves on.

Preface to 2015 Version

This project began when I was trying to understand why Einstein, in his General Theory of Relativity, believed that the universe was limited. (A limited universe is just as difficult to understand as an unlimited one.) First I needed to spend more time studying his Special Theory of Relativity, which was the foundation of the general theory, and so in 1970-71 I immersed myself into the subject matter. After studying it carefully I submitted two slightly different papers to various science publications and publishers in 1971-72. One accepted it, *Philosophy of Science*, and it went through several reviews before it was ultimately turned down for what turned out to be incorrect reasons, as we will explain in this book. Yes, Einstein's theory is believed to be confirmed by most scientists, and thus it is almost apostasy to argue against it as it was to argue against Ptolemy's geocentric theory centuries ago. Yet it was based mostly on a misunderstanding of the nature of light, time and motion. Correcting this misunderstanding could solve problems in physics and other disciplines of science, such as astronomy. Today in 2015, physics and astronomy are boxed into the mathematical and paradoxical aspects of the new physics. This new physics came out of the struggle to understand relativity, light and the idea of aether and whether there was really any absolute reference system.

Since 1972, I continued to collect information on the subject and other relevant information. I knew I could not fight a universal mindset and so I waited until 2014 when I decided to review all my collected information and then rewrite the paper into a book. I simplified my ideas as much as possible so the non-mathematician and the general public could understand. Although Einstein's theory of relativity is believed to be incomprehensible to the general public and to many scientists, it was actually a very simple idea, a brilliant idea, to resolve the apparent enigma of light and the relative motion of all other things.

Please read all the footnotes and the papers in the Appendices, for they add significantly to the understanding of this book. Remember, I simplified this work to make it as easy as possible to understand. Complex mathematics are not needed to see the perceptional mistakes of the train, the track and lightning in the example most often given to explain away simultaneity and time. I welcome all constructive criticism. — WRD, February 22, 2015

Preface to the 2025 Edition

This version simplifies parts of the last book, as well as adds important information about light. For one thing, light is energy and should adhere to the Second Law of Thermodynamics. All our devices that measure light are really *energy* detecting devices. Shifts in energy and velocity manifests itself by changes in the doppler effect. We explain how the various energy rays within the electromagnetic spectrum (EMS) are merged together blinding us to their respective velocities. We present an experiment to detect the various velocities within the EMS by detecting and measuring light as it peeks around the moon during an eclipse of the sun . We also discuss the Aether-Field, which replaces the name "Space," "Field," and "Coordinates of Space and Time" and give hints as to what its essence really is. Can light ever go faster than the artificial limit set for light? It already does, but to see it and measure it we need to know more about what light is, and this is what this book is trying to achieve.

We also continue to critic the many assumptions of science because if these assumptions are wrong, then our various departments in science may also be wrong. Be sure to read the Appendices of this book.

Read our Table of Contents page to see all the subjects we examine.

Questions: What is light? Why is there anything? Is time a thing or just a comparative quality the mind uses to ascertain reality? So many questions.

Remember science is only as good as its honest use of the scientific *method*.

Walter R. Dolen

September, 2025

Table of Contents

Preface to 2015 Version 6

Preface to the 2025 Edition 7

§ 1. Introduction   12

Relativity of Motion 12

Relativity in a Multi-bodied Universe 12

    Background: Why this Paper? 15

My First Paper and E = mc2 15

    E = mc2 Before Einstein..15

§ 2. Premises of Special Relativity   18

Einstein's Premise 1: 18

Einstein's Premise 2: 18

Einstein's 'Correction' to the Classical Principle of Relativity 19

Lorentz Transformation..20

    Lorentz Transformation: "A Gift From Above" 20

Commonsense and Mathematics 21

High Velocity v. Low Velocity 21

    Only When the Velocity is High 22

Science and the New Physics...22

§ 3. Time and Relativity   23

Lorentz and Local Time..24

Traditional Measurement and Unit of Time.........................25

    Universal Time and the Number Base System. 25

    What is Time?...25

    Changing the meaning of words.....................................26

"The Relativity of Simultaneity" According to Einstein.........................26

My Comment..27

§ 4. The Nature of Light   28

Interrelationship of Velocity, Wave Length, Frequency, and Energy 28

Wave-Particle Duality 28

Light's Wave Formula..29

Light's Energy, Frequency Relationship...............................29

Humans' Detection of Light and Color...30

Instrumental Detection of Light..30

Sources of Light..31

Cosmic Objects & Superluminal Motion 31

Great Distances in Space...32

Theory of Redshift Phenomenon and Distance to Galaxies.................33

Light's Velocity Variability 34

Does Light's Velocity Vary in a Gyroscope? 35

Experiment Shows Relativity of Light's Velocity................................37

Doppler Effect..38

Yes, Light's velocity is Independent of source 40

Doppler Effect and Relative Velocity 41

§ 5. Confusion Between the Nature of Light and Velocity   42

Figure 9: Graphic by Walter R. Dolen..44

Electromagnetic Velocities/Energies appear to merge........................44

Experiment to Ascertain the Various Velocities of Light 45

§ 6. An Aether-Field in the Universe   46

Acceleration and Inertial..46

Aether-Field 47

Conclusions From Our Study of Einstein's Special Theory of Relativity (STR)   48

§ 7. Appendices — Assumptions of Science   51

• Heisenberg and Dirac tell their stories about the guessing game in their own words, from their 1968 lectures, as published in Abdus Salam's book..51

• Basic Assumptions of Quantum Theory are Arbitrary 51

• Scientists Observe Through Their Theories........................51

• Constancy in Science Was Often Only Assumptions 51

Appendix 1: Relativity of Mass?   52

Appendix 2: Contradiction of Clocks   55

Appendix 3: Fourth Dimension   59

Appendix 4: General Theory, Black Holes & Antimatter   60

Appendix 5: GPS and Relativity   62

Appendix 6: Mathematics and Reality   63

Math is a Language 64

Arithmetic and Real Things...64

Arbitrary Rules in Math 65

Math Rules 65

Imaginary Numbers 66

Multiple answers for the same equation..................................67

Misusage of the language of words: 68

    No man ever steps in the same river twice? 68

Misusage of the language of numbers: 68

    Zeno: Motion Impossible?...68

Appendix 7: Science is a Guessing Game   70

Guessing in Mathematics by Famous Scientists....................70

Following Taken from Heisenberg's 1968 Lecture   72

Phenomenological Theory..72

Bohr's Conjecture 72

Einstein on Theory and Observation 75

Rigorous and Dirty Mathematics..76

Abandoning Old Concepts 77

Quantum Theory Understood 77

Electrons and the Nucleus 78

Changing the Outlook of Atomic Physics.................................79

Pair Creation..79

Pauli's Critical Acumen 80

The Following Taken From Paul Dirac's 1968 Lecture: 81

Cosmological Speculation 81

Impact of Relativity..81

On the Wrong Track? 82

Difficulty in Quantum Electrodynamics..................................83

Basic Assumptions of Quantum Theory are Arbitrary   84

All Electrons Are Identical?...84

Scientists Observe Through Their Theories   85

Einstein put it this way: 85

Hanson said it this way: 85

Constancy in Science Was Often Only Assumptions   85

Light's Velocity in a Vacuum Always the Same? 85

Constant Decay Rates? 86

Constant Atomic Clock Rates? 87

My Conclusion about the Guessing Game   87

Appendix 8: Why is There Anything   88

Why is there anything in the Universe. Why? Can something come from nothing. 88

Recent Theories (as of 2025) on the Universe's Origin.........................90

Flashlight's/Searchlight's Diminishing Power: Energy = Distance ...91

Appendix 9: Original Papers   92

About the Author   99

Einstein: Light, Time, and Relativity

This paper examines: (1) Special Theory of Relativity's premises and paradoxes; (2) time, relative and constant velocities; (3) the nature of light and its velocity; (4) Gravity, what is it?; (5) and the theory's perceptional confusion.

§ 1. Introduction

Before we can ascertain Special Theory of Relativity's uncertain underpinnings we need to understand it. In order to understand the theory we need to present some background and the reason for the development of the Special Theory of Relativity (STR).

Relativity of Motion

Among other things, any theory of relativity deals with methods of translating the three dimensional coordinates (x, y, z) from one coordinate system (trains, cars, stars, planets, et al.) to another when they are moving relative to each other. In normal circumstances you would use the classical (Newtonian/Galilean) transformation equation.[1] For example, in the graphic below, the black car that is moving at a *constant* speed of 30 kilometers per hour (km/h) down a road is also moving at a velocity of 30 km/h *relative* to the road. If a gray car behind him (traveling the same road in the same direction) is moving at a constant 20 km/h, then the black car in front is moving 10 km/h faster relative to the gray car behind (you subtract the two velocities), while moving 30 km/h relative and constant to the road. If a white car (in front of the black car) is moving at a constant 30 km/h in the opposite direction on the same road, then the relative speed of the black car and the white car is 60 km/h towards each other (you add the two velocities together), yet each car is at a constant 30 km/h relative to the road.

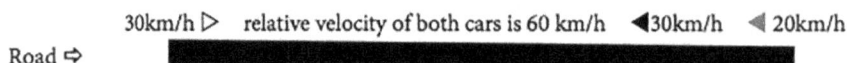

30km/h ▷   relative velocity of both cars is 60 km/h   ◀30km/h   ◀ 20km/h

Road ⇨

While all the cars are in motion relative to the road and to each other on the earth, the Earth and the cars also are moving (orbiting) through the solar system at about 30 km/s (kilometers per second), while our sun and solar system seem to be orbiting through the Milky Way galaxy at about 220 km/s, and the Milky Way galaxy may also be moving at another speed relative to another reference system.

Relativity in a Multi-bodied Universe

If there were *only* two bodies in the universe, it would be impossible to know which body was moving, or not moving, or at what speed. However,

---

[1] x' = x - vt ; (2) t' = t — where v is the relative velocity and x, t is one frame of reference and x', t' is the other frame of reference. Time (t or t') is universal time.

there are more than two bodies in our universe. In a universe of multiple bodies one can construct a system (*e.g.* using GPS) to ascertain which bodies are moving or not moving relative to each other. For example, Ptolemy's system[2] incorrectly had the Earth apparently stationary while the sun and stars were revolving around the Earth each day. Logic and physical evidence through observation and comparison negated this mindset. Through scientific methods, we today can approximate distances in space, relative velocities of stars and planets, etc. It is not correct to state that we cannot ascertain, through comparison, which bodies are moving in any two reference systems[3] since we exist in a multiple-body universe — not a two-body universe. Even Einstein knew one could use the entire universe as a reference system. In 1917 he wrote:

> The most important fact that we draw from experience as to the distribution of matter is that the relative velocities of the stars are very small as compared with the velocity of light. So I think that for the present we may base our reason upon the following approximative assumption. **There is a system of reference** relatively to which matter may be looked upon as being permanently at rest. [4]

In context, Einstein was using the whole universe as a reference system. Einstein later in his life in 1950 said the same thing when writing about acceleration — which needs to be referenced to something:

> "Inertia resists acceleration, but acceleration relative to what? Within the frame of classical mechanics the only answer is: Inertia resists acceleration *relative to space.*"[5]

But Einstein in this paper seems to say that acceleration in his General Theory is accelerated relative to the "continuous field" represented in his theory by the **coordinates of space and time**:

> "According to general relativity, the concept of space detached from any physical content does not exist. The physical reality of space is represented by a field whose components are continuous functions of four independent variables — coordinates of space and time."

---

[2] *Ptolemy's Almagest*, translated into English and annotated by G. J. Toomer, 1984, Publ: Springer-Verlag. Ptolemy's system used mathematics to successfully predict the future movement of the stars and planets, and contrary to popular accounts, the *Almagest* was a scientific document, not a religious document.

[3] Only if the occupant of a train is disoriented can he say the train is stationary and the embankment is moving. Scientific knowledge through a GPS or other observation would clarify the situation.

[4] A. Einstein, "Cosmological Considerations on the General Theory of Relativity," IX, § 3, 1917 found in — A. Einstein, *The Principle of Relativity: A Collection of Original Papers on the Special and General Theory of Relativity*, 1952, Dover edition. Emphasis added.

[5] A. Einstein, "On the Generalized Theory," *Scientific American*, April 1950 [reprint], p. 5. Emphasis is in text.

Therefore, Einstein seems to say that acceleration is relative to the coordinates of space and time or that is — the whole universe of his space and time.

Geoffrey Builder of the School of Physics at the University of Sydney, stated it this way (as Einstein did that the Universe can be used as a reference system), when arguing for the ether[6] hypothesis:

> The relative retardation of clocks, predicted by the restricted theory of relativity demands our recognition of the causal significance of absolute velocities. This demand is also implied by the relativistic equations of electrodynamics and even by the formulation of the restricted theory itself. **The observable effects of absolute accelerations and of absolute velocities must be ascribed to interaction of bodies and physical systems with some absolute inertial system. We have no alternative but to identify this absolute system with the universe**. Thus, in the context of physics, absolute motion must be understood to mean motion relative to the universe....[p. 279, my emphasis]

> The absolute character of acceleration forces us, like Newton, to postulate some absolute universal system relative to which bodies and systems are accelerated. Moreover, the possibility of detecting absolute acceleration by its dynamical effects forces us to ascribe these effects to interaction between the bodies affected and this absolute system. In particular, it is necessary to ascribe the inertia of bodies to such interaction. The necessity for this has been argued fully by Mach, Einstein, and others. [p.287]

> There is no feasible alternative to supposing that **this absolute system is the universe as a whole**, or else something universal which is an integral and essential part of the universe. [p. 289, my emphasis][7]

Although Einstein discounted the need for ether[8] when stating his relativity theory, he did not discount the possibility of ether's existence:

> More careful reflection teaches us, however, that the special theory of relativity does not compel us to deny ether. We may assume the existence of an ether; only we must give up ascribing a definite state of motion to it....[9]

We will write about this ether (Aether-Field) later in this paper.

---

[6] In the late 19th century, luminiferous aether, or ether, meant the "light-bearing aether." It was believed to be the medium for the propagation of light.

[7] G. Builder, "Ether and Relativity," *Australian Journal of Physics*, vol. 11, p. 279, 1958.

[8] "The introduction of a 'luminiferous ether' will prove to be superfluous inasmuch as the view here to be developed will not require an 'absolutely stationary space.'" Published in German in *Annalen der Physik.* **17**:891 (1905).

[9] Albert Einstein, *Sidelights on Relativity,* an address delivered on May 5th, 1920, in the University of Leyden. Amazon sells a copy of this on their website as of 2014.

Background: Why this Paper?

In 1971, after reading about Einstein's Special and General theories on relativity, something struck me about them besides their paradoxes. At first, I took it for granted that the theories were correct since I was assured by knowledgeable books and scientists. I was especially interested by his contention about the shape and extent of the universe, which he derived from his 1916 General Theory of Relativity:

> "As I have shown in the previous paper, the general theory of relativity **requires that the universe be spatially finite.**"[10]

From this assertion I took on the task to study his theories in depth. First I studied the Special Theory of Relativity (STR) from which the General Theory came.

My First Paper and E = mc²

I wrote a paper after my study *(The Totality of Motion's Relativity,* 1971-72) and sent it out to various science publications, including the *Philosophy of Science*, which accepted my study and put it through its normal peer review. It was finally turned down at the last stage of review by the last referee because he believed that the atomic power/bomb/E=mc² seemed to confirm Einstein's theory. This was untenable because Einstein's theory had almost nothing to do with the atomic bomb.[11] The only thing he contributed to the development of the atomic bomb was the $E = mc^2$ formula, which is called the *mass-energy equivalence.*[12] This equation suggested that there was energy in mass. Although everyone knew this already because coal or wood has energy that is released when burned, according to the equation it is the *extraordinary* amount of energy in mass that is pertinent.

E = mc² *Before* Einstein

Others *before* Einstein also speculated that there could be an enormous amount of energy in mass. As far back as the 1700s we see what Sir Isaac Newton said, "gross bodies and light are convertible into one another."[13] In context he was saying, the energy of light is released from heated objects, and the energy of light can be absorbed into an object:

---

[10] A. Einstein, *The Principle of Relativity: A Collection of Original Papers on the Special and General Theory of Relativity*, 1952, Dover edition, p. 193. My emphasis.

[11] Richard Rhodes, *The Making of the Atomic Bomb*, pp. 305-307.

[12] Einstein, A. (1905), "Ist die Trägheit eines Körpers von seinem Energieinhalt abhängig?" *Annalen der Physik* **18** (13): 639–643. In English the title is, *"Does the inertia of a body depend upon its energy content?"*(1905) In this paper, Einstein used V to mean the speed of light in a vacuum and L to mean the energy lost by a body in the form of radiation. Consequently, the equation $E = mc^2$ was not originally written as a formula but as a sentence in German saying that *if a body gives off the energy* L *in the form of radiation, its mass diminishes by* L/V² which, using the more modern E instead of L to denote energy, may be trivially rewritten as "E = mc²". A remark placed above it informed that the equation was approximated by neglecting "magnitudes of fourth and higher orders" of a series expansion. English translation here — fourmilab.ch/etexts/einstein/E_mc2/www/

[13] Sir Isaac Newton, *Opticks*, 4th ed (1730), p. 374, Under "Quest. 30."

> For all fix'd Bodies being heated emit Light so long as they continue sufficiently hot, and Light mutually stops in Bodies as often as its Rays strike upon their Parts, as we shew'd above.[14]

During the nineteenth century, in various theories pertaining to ether, attempts were made to manifest that mass and energy were interrelated.[15] In 1873, Nikolay Umov showed the relationship between mass and energy in his formula $E = kmc^2$. He is credited with being the first scientist to indicate the interrelationship between mass and energy.[16] Also written works by Samuel Tolver Preston (1875)[17] and Olinto De Pretto (1903)[18] showed the mass-energy relationship. Both Preston and De Pretto, following Georges-Louis Le Sage (1748), believed that the universe was filled with tiny particles moving at the speed of light, with each particle having a kinetic energy of mc².

We see then that Einstein was *not* the first to suggest the idea of the equivalence of mass and energy nor was he the first to use the equation $E = mc^2$ for energy.[19] The equation was first published in the scientific magazine *Atte* by Olinto De Pretto in 1903 (two years before Einstein) and then republished by the Royal Science Institute of Veneto in 1904. Although Einstein was not the first, he greatly facilitated the idea for the potential of releasing a vast amount of energy from mass through the popularization of his theory in the world's press after the 1919 eclipse. I ask though, what does the *velocity* of light have to do with mass? Or rather, what does the velocity of light multiplied by itself have to do with mass or energy? Why multiplied by itself? Why not multiplied to the third power, or to the fourth power, or $E = mc^{2.5}$? The equation is simply a guess as Niels Bohr[20] guessed his mathematics for the electron orbits of the elements in his periodic table,

---

[14] Ibid., p. 374.

[15] Helge Kragh, "Fin-de-Siècle Physics: A World Picture in Flux" in *Quantum Generations: A History of Physics in the Twentieth Century*, Princeton, NJ: Princeton University Press, 1999.

[16] *Умов Н. А. Избранные сочинени N.A.* [Umov Selected Works— 1950]

[17] Preston, S. T., *Physics of the Ether*, E. & F. N. Spon, London, (1875).

[18] De Pretto, O. *Reale Instituto Veneto Di Scienze, Lettere Ed Arti*, LXIII, II, 439–500, reprinted in — Umberto Bartocci, *Albert Einstein e Olinto De Pretto—La vera storia della formula più famosa del mondo*, editore Andromeda, Bologna, 1999; see also "Who invented Relativity." — http://www.mathpages.com/rr/s8-08/8-08.htm [accessed Aug., 2025]

[19] For more detail see Christian Bizouard's paper (accessed Aug., 2025) at http://www.annales.org/archives/x/poincaBizouard.pdf and other articles on the Internet such as

Ajay Sharma, "Origin and Escalation of the Mass-Energy Equation $\Delta E = \Delta mc^2$," The General Science Journal, accessed Aug. 2025 at https://www.gsjournal.net/Science-Journals/Research-Papers-Relativity-Theory/Download/2088

[20] Niels Bohr was a Danish physicist who contributed to the understanding of atomic structure and quantum theory. He won the Nobel Prize in Physics in 1922.

and as other famous scientists guessed their famous findings.[21] There is no logical connection between *velocity* and mass let alone between the velocity of *light* and mass. The fact that atomic bombs have only released a very *small* (less than ½ percent of the) amount of energy predicted by Einstein's equation is one huge piece of evidence against the equation:

> Fission reactions turn only about .09 percent of atomic mass into energy; Fusion reactions turn about .4 percent of mass into energy.[22]

Instead of the power released by atomic power/weapons supporting the $E = mc^2$ equation, it is actually pointing a laser beam at the shortcoming of the equation. Also see the Appendix 1, "The Relativity of Mass?," for a further discussion on $E = mc^2$.

Let's look at Einstein's premises to better understand the theory.

---

[21] *Unification of Fundamental Forces: The First of the 1988 Dirac Memorial Lectures* by Abdus Salam [Cambridge University Press, Cambridge, GB], 1990 pp. 90-121. See Appendix 7.

[22] William J. Broad, *New York Times*, Sept 15, 1985, p. 17, Sunday "Science" Section.

§ 2. Premises of Special Relativity

Einstein's theory is based on a foundation, and from that foundation, the mathematical equations were deduced. Because of this his equations are no better than his premises. There are two basic premises in his paper on relativity. Premises 1 and 2 below are the original wordings translated into English from his 1905 paper, *On The Electrodynamics of Moving Bodies.*[23]

Einstein's Premise 1:

> **The same laws of electrodynamics and optics will be valid for all frames of reference for which the equations of mechanics hold good.** We will raise this conjecture (the purport of which will hereafter be called the "Principle of Relativity") to the status of a postulate.

Or Premise 1 as restated by Einstein in 1950:

> The "principle of relativity" in its widest sense is contained in the statement: The totality of physical phenomena is of such a character that it gives no basis for the introduction of the concept of "absolute motion"; or shorter but less precise: **There is no absolute motion.**[24] [Thus all motion is relative.]

Einstein's **Premise 1** is generally the same principle of relativity as put forth by Newton in 1687. "The motion of bodies included in a given space are the same among themselves, whether that space is at rest or moves uniformly forward in a straight line."[25] This is known as the classical or Newtonian or Galilean relativity principle. Einstein's first premise is generally the same principle as the classical theory of relativity.

Einstein's Premise 2:

> And [he] also introduce another postulate, which is only apparently irreconcilable with the former, namely **that light is always propagated in empty space with a definite velocity c which is independent of the state of motion of the emitting body**. These two postulates suffice for the attainment of a simple and consistent theory of the electrodynamics of moving bodies based on Maxwell's theory for stationary bodies. The introduction of a "luminiferous

---

[23] Published in German in *Annalen der Physik.* **17**:891, 1905; Emphasis added. English translation — www.fourmilab.ch/etexts/einstein/specrel/www/ [Link good as of Aug, 2025]

[24] Albert Einstein, *Out of My Later Years* (New York: Philosophical Library, Inc., 1950), p. 41. Emphasis added.

[25] "The motion of bodies included in a given space are the same among themselves, whether that space is at rest or moves uniformly forward in a straight line." [Sir Isaac Newton, *Mathematical Principles of Natural Philosophy*, Andrew Motte, trans. (1729 ed.), Florian Cajori, revised trans. (Berkeley: University of Calif Press, 1946), p. 20; and note Lincoln Barnett, *The Universe and Dr. Einstein* (New York: William Sleane Assoc., 1957), pp. 31-32.] Or, "The motions of bodies included in a given space are the same among themselves, whether that space is at rest, or moves uniformly forwards in a right [straight] line without any circular motion." [Newton's Corollary V, from Axioms, or Laws of Motion, Translated by Andrew Motte, 1846]

ether" will prove to be superfluous inasmuch as the view here to be developed will not require an "absolutely stationary space."[26]

And note Einstein's Premise 2 as restated by the physicist Richard C. Tolman in 1910:

> ***The constant velocity of light.*** The velocity of light as measured by any observer will be found to have one and only one value (a "universal constant") under all circumstances: that is, regardless of the motion of the source relative to the observer.[27]

Einstein's 'Correction' to the Classical Principle of Relativity

More specifically, there are three assumptions of the classical (Newtonian/Galilean) principle of relativity:

> (1) There are inertial systems[28] (closed systems of reference) in which the laws of physics are identical.
>
> (2) If one system is inertial, any system of reference's velocity which is constant relative to the first one is also inertial.
>
> (3) The transcription of space and time data from one inertial system to another has to be done according to the **Galilean transformations**.[29]

Premise 1 of special relativity was deduced from these three assumptions just mentioned, and the fact that a motionless cosmic aether had never been proven.[30] But Premise 2 of special relativity, which was deduced from Michelson-Morley experiments and other astronomical observations,[31] is contrary to Einstein's Premise 1 because the velocity/motion of light does not appear to be relative to anything else: all motion is relative except for the velocity of light in a vacuum.

It was the seeming uniqueness of light's velocity that motivated Einstein to construct his theory. In Einstein's own words:

---

[26] Published in German in *Annalen der Physik.* **17**:891, 1905; English translation — www.fourmilab.ch/etexts/einstein/specrel/www/ Emphasis added. Link good as of August, 2025.

[27] Richard C. Tolman, *Phys. Rev.* **31**, 27 (1910); see "Relativity," *Ency. Internationali* vol. **15**, p. 356 (1970); etc. [My bold emphasis, his italic emphasis]

[28] In classical physics and special relativity, an inertial frame of reference (also called an inertial space or a Galilean reference frame) is a frame of reference in which objects exhibit inertia: they remain at rest or in uniform motion relative to the frame until acted upon by external forces.

[29] "Relativity," *Ency. Brit.* Vol. 19, pp. 96-97 (1969).

[30] Otherwise all motion would be relative to a possible stationary aether, as well as being relative to each other. The Michelson-Morley experiment and other like it convinced scientists of this, although the. Majorana experiment (Phil. Mag. S. 6. Vol. 37. No. 217. Jan. 1919) seemed to counter Michelson-Morley's finding.

[31] Binary stars, etc.

"Thus the special theory of relativity does not depart from classical mechanics through the postulate of relativity [Premise 1], but through the postulate of the constancy of the velocity of light in vacuo."[32]

Lorentz Transformation

From the apparent contradiction between Premise 1 and 2, Einstein borrowed formulas which mathematically incorporated both premises. The special theory of relativity "corrects" the Galilean transformation in Assumption 3 above by changing it as follows:

> (3) The transcription of space and time data from one inertial system to another is done according to the **Lorentz transformation**.

This transformation system is called the Lorentz transformation instead of the Einstein transformation system because the Dutch physicist, Hendrik Antoon Lorentz, independently constructed this system before Einstein. The difference between the two transformation formulas is as follows. In the Galilean equations time, space (length) and mass are fixed and all velocities are relative. In the Lorentz equations time, space and mass are relative, but the speed of light is constant.[33] Therefore in Einstein's theory the only constant is the velocity of light, while time, space (length) and mass can change depending on the relative speed of the body compared to the speed of light. This is merely a phenomenon of Lorentz's equation. It only exists in the math. It is impossible to get a picturable example that depicts all the different "local time" possibilities: "Lorentz's theory had to introduce a special measure of length and time for **every** moving system."[34] The so-called "proof" of clocks running slow has mostly to do with atomic clocks, but this ignores the fact that these clocks are affected by gravity, temperature, solar activity, Earth-Sun distance, or other exterior forces.[35] Computer online applications to figure the *supposed* length contraction, time dilation and mass relativity in STR can be found online.[36]

Lorentz Transformation: "A Gift From Above"

The famous mathematician Hermann Minkowski wrote this in 1908:

---

[32] A. Einstein, *The Principle of Relativity: A Collection of Original Papers on the Special and General Theory of Relativity*, Dover edition 1952, p. 111. (Originally published by Methuen and Company 1923.)

[33] The irony here is that the Lorentz transformation was formulated to help support the aether theory, while Einstein used the same formula to de-emphasize the aether theory.

[34] See § 3 under, "Lorentz and Local Time?" Emphasis added.

[35] G. T. Emery, "Perturbation of Nuclear Decay Rates," *Annual Review of Nuclear Science*, Vol. 22, 1972, pp. 165-202; "The Mystery of varying nuclear decay," *Physics World*, Oct 2, 2008; Jere H. Jenkins, Ephraim Fischbach, "Perturbation of Nuclear Decay Rates During the Solar Flare of 13 December 2006," *Astroparticle Phys.* 31:407-411, 2009.

[36] In Aug. 2025 such apps were active on: http://hyperphysics.phy-astr.gsu.edu/hbase/Relativ/tdil.html

According to Lorentz any moving body must have undergone a contraction in the direction of its motion.... This hypothesis sounds extremely fantastical, for the contraction is not to be looked upon as a consequence of resistances in the ether, or anything of that kind, but simply as a gift from above...."[37]

If it were not for the existence of the Lorentz Transformation formula, there would not be an Einstein's relativity theory today, so it was like a gift from on high. It made it possible mathematically to save the idea of the constant velocity of light, but in its wake it attacked logic and common sense.

Commonsense and Mathematics

Before Einstein, physics was called the physics of commonsense because it was perceivable and mechanically explainable. But after Einstein, physics moved beyond commonsense to where only the mathematician could perceive the cosmos as it is. And so in A. d'Abro's history about the rise of the New Physics he states:

The exalted laws of logic, long assumed to be the laws of the human mind, thus appear to be mere empirical laws, subject to revision as the field of our mathematical experience is widened.[38]

Also Morris Kline (May 1, 1908 – June 10, 1992) who was a Professor of Mathematics (NYU) and a writer on history, philosophy and math adds in his *Mathematics and the Physical World*:

Indeed, physical science has reached the curious state in which the firm bold essence of its best theories is entirely mathematical whereas the physical meanings are vague, incomplete, and in some instances even self-contradictory. **Science has become a collection of mathematical theories adorned with a few physical facts.**[39]

Read Appendix 6 for more information on mathematics and reality.

High Velocity v. Low Velocity

The classical or Galilean transformation is the commonsense transformation formula because it agrees with our everyday experience. Length and time do not change. Einstein's theory, in cases where the velocity of mass nears the speed of light, is not a commonsense theory since time, space and mass are made relative to the speed of light. Again from D'Abro we see this:

Many of the conclusions derived from the theory of relativity conflict strongly with the dictates of commonsense.... **But the theory of relativity shows that the dictates of commonsense are**

---

[37] H. Minkowski, "Space and Time," 1908, found in — A. Einstein, *The Principle of Relativity: A Collection of Original Papers on the Special and General Theory of Relativity*, Dover edition 1952, p. 81. (Originally published by Methuen and Company 1923.)

[38] A. D'Abro, *The Rise of the New Physics, Vol. One*, Dover, p. 43.

[39] Morris Kline, *Mathematics and the Physical World*, 1959, p. vii. Emphasis added.

contradicted *only* when the velocities compounded are extremely high.... For low velocities, the relativistic rule of composition tends to differ less and less from the classical rule, which is supported by commonsense.[40]

Only When the Velocity is High

This is important to understand: *only* when the velocity of a body is very high does the Special Theory of Relativity vary from the classical system — the system of commonsense. The equation was set up this way: at low velocities, almost every "proof" of Special Theory of Relativity can also prove the classical system because at low velocities there are no differences. The difference between the measurement of classical or Einstein' relativity in the Global Positioning System (GPS) is less than one centimeter.[41]

Science and the New Physics

In part, science's present status stems from the fact that it broke the metaphysic and mystical arguments used to keep the second-century geocentric-mathematical system of Ptolemy in place, and replaced it with the heliocentric system, which was put together through the work of Copernicus, Galileo, Kepler and others. This heliocentric system replaced the old system with a better description of the solar system by using the Occam's razor: the fewer the assumptions the better the hypothesis. The more accurate results of the sun-centered system, and the information from satellites sent within our solar system helped to firmly establish the new system as the true picture of our solar system. But today's science has turned its back on the science of observation and the Occam's razor and replaced it with mathematical and creative metaphysics. Is it possible to go back to observational science?

---

[40] A. D'Abro, *The Rise of The New Physics*, Volume Two, Dover, 1952 (originally pub 1939 by D. Van Nostrand Co.), p. 438. Emphasis added.

[41] see Appendix 5 in this book

§ 3. Time and Relativity

This section is very important because it deals with the very foundation of Einstein's Special Theory of Relativity.[42] Therefore I will give several examples to clearly make my point.

In Einstein's 1905 relativity paper he used a thought-experiment to manifest what he believed was the relativity of simultaneous events.[43] His initial paper on this subject was confusing and not as clear as his later renderings. **Einstein's questioning of simultaneity was his way of attacking the idea of time being a constant**. The doubts about the nature of time were positioned into the minds of physicists by the 1895 book by H. G. Wells called the *Time Machine*:

> [Time Traveller:] There are really four dimensions, three which we call the three planes of Space, and a fourth, Time. [...] Our mental existences, which are immaterial and have no dimensions, are passing along the Time-Dimension with a uniform velocity from the cradle to the grave. [...]
>
> "But the great difficulty is this," interrupted the Psychologist. "You can move about in all directions of Space, but you cannot move about in Time." [...]
>
> [Time Traveller:] That is the germ of my great discovery. But you are wrong to say that we cannot move about in Time." [44] [etc.]

And the doubts about time were aided by such papers as the 1898 paper by the eminent mathematician and theoretical physicist, Henri Poincaré, "The Measure of Time." In this paper Poincaré pointed out the apparent inconsistency of time — the "approximation" of time:

> To measure time they use the pendulum and they suppose by definition that all the beats of this pendulum are of equal duration. But this is only a first approximation; the temperature, the resistance of the air, the barometric pressure, make the pace of the pendulum vary. If we could escape these sources of error, we should obtain a much closer approximation, but it would still be only an approximation....
>
> In fact, the best chronometers must be corrected from time to time, and the corrections are made by the aid of astronomic observations;

---

[42] We are quite aware that Einstein did not create the theory by himself; he was merely the one credited with it by mass media and some of his contemporary scientists. See the book, *How Einstein Ruined Physics* (2011), by Roger Schlafly for info about where and from whom Einstein got his ideas. The title of this book is unfortunate since most of the content was about from whom Einstein borrowed his ideas, not on how he "ruined" physics.

[43] A. Einstein, *The Principle of Relativity: A Collection of Original Papers on the Special and General Theory of Relativity*, 1952 Dover edition, pp. 38ff.

[44] Wells, H. G. (Herbert George). *The Time Machine* (p. 2, 5). Kindle Edition.

arrangements are made so that the sidereal clock marks the same hour when the same star passes the meridian.[45]

In this same 1898 paper about time, Poincaré tries to darken the idea of universal time by manifesting its approximation:

> "One of the circumstances of any phenomenon is the velocity of the earth's rotation; if this velocity of rotation varies, it constitutes in the reproduction of this phenomenon a circumstance which no longer remains the same....Our definition [of time] is therefore not yet satisfactory."[46]

He darkens further the idea of time and simultaneity by concluding:

> To conclude: We have not a direct intuition of simultaneity, nor of the equality of two durations. If we think we have this intuition, this is an illusion. We replace it by the aid of certain rules which we apply almost always without taking count of them.... We therefore choose these rules, not because they are true, but because they are the most convenient, and we may recapitulate them as follows: "The simultaneity of two events, or the order of their succession, the equality of two durations, are to be so defined that the enunciation of the natural laws may be as simple as possible. In other words, all these rules, all these definitions are only the fruit of an unconscious opportunism."[47]

Lorentz and Local Time

The idea of constancy of time was also attacked by Hendrik Antoon Lorentz who introduced "local time" in 1895: "We can therefore call this variable the local time [Ortszeit] of this point, in contrast to the general time t."[48]

Einstein, through Lorentz's original idea, introduced non-universal time or "local" time in order to further his relativity theory. Without local time there could never be a constant velocity of light. In the words of the famous physicist, Max Born:

---

[45] In French: Poincaré, Henri (1898), "La mesure du temps", *Revue de métaphysique et de morale* 6:1-13; English translation: *Henri Poincaré,* (1913), "The Measure of Time", *The Foundations of Science (The Value of Science), New York: Science Press, pp. 222-234.* English translation: *https://en.wikisource.org/wiki/The_Measure_of_Time*, p. 224-225. [Accessed Aug., 2025]

[46] Ibid., p. 227.

[47] Ibid., p. 234.

[48] H.A. Lorentz, "Attempt of a Theory of Electrical and Optical Phenomena in Moving Bodies," p. 50 of the English translation from "Versuch einer Theorie der elektrischen und optischen Er-scheinungen in bewegten K'orpern," Leiden: E.J. Brill (1895). pp. 1-138. German text can be accessed in Google Books as of Aug. 2025; English translation by Wikisource — https://en.wikisource.org/wiki/ Translation:Attempt_of_a_Theory_of_Electrical_and_Optical_Phenomena_in_Moving_Bodies

A new time measure must be used in a system which is moving uniformly. He called this time, which differs from system to system, "local time." [...] To maintain the law of constancy of the propagation of light, Lorentz's theory had to introduce a special measure of length and time for every moving system.[49]

Traditional Measurement and Unit of Time

How was time measured up to this point in history by peoples and scientists? Time was defined as James Clerk Maxwell defined it in his *A Treatise on Electricity and Magnetism*, after the heading "The Three Fundamental Units" [Length, Time, Mass]:

> *Time.* The standard unit of time in all civilized countries is deduced from the time of rotation of the earth about its axis....The unit of time adopted in all physical researches is one second of mean time.[50]

In context, Maxwell was speaking about fundamental units: length, time and mass. For most of our history the reference system for time was the daily rotation of the earth about its axis in relationship, of course, to cosmic bodies.[51] This method was and is not perfectly constant because the rotation of the earth seems to vary minutely over the months and years. But this does not matter as long as all scientists agree on a relatively constant reference system to measure time.

Universal Time and the Number Base System.

Agreeing to use the concept of universal time is similar to agreeing to use the same **base** system for our math. If we used the base 8 system, instead of the decimal base (base 10), all our equations would produce different results because **23** in a base 8 system is **19 (2x8+3=19)**in a base 10 system **23 is 23 (2x10+3=23)**. Therefore, all scientists today use the decimal base system (base 10), for the same reason all scientists used universal time *before* the new physics that Einstein helped to create.

What is Time?

Knowing and using a universal unit of time is not the same as knowing what is time. Time was never a physical object or an intrinsic aspect of the universe. It was never the fourth dimension of the universe as in Einstein's General Theory of Relativity. Time was and is merely a comparative mental quality or concept used to perceive the ideas of velocity, age, history, movement, simultaneity or duration in our universe. Time is an abstract idea. When time refers to duration, it refers to the movement or behavior of real objects in the universe as measured by a reference system (e.g. the second, hour, day, month, year, century, millennium, etc). Our mind uses the

---

[49] Max Born, *Einstein's Theory of Relativity*, pp. 222 & 225, Dover 1965 Ed. First pub. 1924.

[50] James Clerk Maxwell, *A Treatise on Electricity and Magnetism*, 3rd ed., quote taken from Dover 1954 reprint edition, Vol. 1, p. 3. First edition published in 1873.

[51] Except many thought the earth was motionless in space and the cosmic bodies revolved around the earth each day.

concept of time to perceive the world around us, but time itself does not exist in the universe as a separate entity. Without using universal time we had chaos up until a couple of centuries ago because all nations used different ways of ascertaining time, dates and reigns of kings. The reason the study of chronology was so mixed up is because most nations had their own methods of chronology — telling time. The reason physics hasn't advance in the last few decades is because they are affected by the confusion of Relativity and Quantum physics. Keep reading.

Changing the meaning of words

Universal time was used to measure aspects of the universe for the same reason we agree to use the same base system in mathematics— to make it possible to communicate with each other. Einstein's theory throws the non-relative and universal units of time and length to the wind and instead replaces them with the constant velocity of light. The fact is, **if you do not have a consistent measurement of time and length, you cannot even measure the velocity of light.** In other words, you cannot ascertain Einstein's only constant (light's velocity) without a non-relative measurement of time and length.

We will see in this paper that simultaneity may seem relative only if you do not use the same time and length, only if you do not have all the facts and only if one thinks light's motion is unique in comparison to **all** other types of movements.

"The Relativity of Simultaneity" According to Einstein

In 1916 in his book *Relativity: The Special and General Theory*, Einstein wrote again about "The Relativity of Simultaneity." He first mentioned this subject in his famous 1905 paper on relativity. In his first paper he did not produce a graphic. In the 1916 rendition (as translated into English in 1920) Einstein included a graphic and wrote the following:

[Note:His example is difficult to perceive, not only because of the way he worded it (or maybe because of the translation) but because he got it wrong.]

> We suppose a very long train travelling [sic] along the rails with the constant velocity *v* and in the direction indicated in Fig. 1 [next page]. People travelling [sic] in this train will with advantage use the train as a rigid reference-body (co-ordinate system); they regard all events in reference to the train. Then every event which takes place along the line also takes place at a particular point of the train. **Also the definition of simultaneity can be given relative to the train in exactly the same way as with respect to the embankment.** As a natural consequence, however, the following question arises:
>
> **Are two events (e.g. the two strikes of lightning *A* and *B*) which are simultaneous *with reference to the railway embankment* also simultaneous *relatively to the train*? We shall show directly that the answer must be in the negative.**
>
> When we say that the lightning strikes *A* and *B* are simultaneous with respect to the embankment, we mean: the rays of light emitted

at the places *A* and *B*, where the lightning occurs, meet each other at the mid-point **M** of the length *A* —> *B* of the embankment. But the events *A* and *B* also correspond to positions *A* and *B* on the train. Let **M'** be the mid-point of the distance *A* —> *B* on the travelling [sic] train. Just when the flashes of lightning occur,[52] this point **M'** naturally coincides with the point **M**, but it moves towards the right in the diagram with the velocity v of the train. If an observer sitting in the position **M'** in the train did not possess this velocity, then he would remain permanently at **M**, and the light rays emitted by the flashes of lightning *A* and *B* would reach him simultaneously, i.e. they would meet just where he is situated. Now in reality (considered with reference to the railway embankment) he is hastening towards the beam of light coming from *B*, whilst he [**M'**] is riding on ahead of the beam of light coming from *A*. Hence the observer [**M'**] will see the beam of light emitted from *B* earlier than he [**M'**] will see that emitted from *A*. Observers who take the railway train as their reference-body must therefore come to the conclusion that the lightning flash *B* took place earlier than the lightning flash *A*. We thus arrive at the important result:

Events which are simultaneous with reference to the embankment are not simultaneous with respect to the train, and *vice versa* (relativity of simultaneity). **Every reference-body (co-ordinate system) has its own particular time**; unless we are told the reference-body to which the statement of time refers, there is no meaning in a statement of the time of an event.[53]

Figure 1

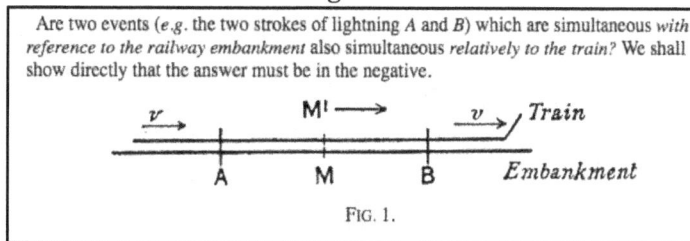

Are two events (*e.g.* the two strokes of lightning *A* and *B*) which are simultaneous *with reference to the railway embankment* also simultaneous *relatively to the train?* We shall show directly that the answer must be in the negative.

FIG. 1.

My Comment

Actually, **M'**, *if* the train has windows in each direction, *and if he is able to see in both directions at once,* then **M'** will NOT see the **B** flash before the **A** flash. What he will see, if he uses an optical instruments that detects the doppler effect in both directions, is that the light he detects from both **A** and **B** reach him simultaneously, but the light from **A** has a red shift doppler effect and the light from **B** has a blue shift effect. Why would M' see the doppler effect instead of the light from B first?

---

[52] As judged from the embankment. [This footnote is in the paper]

[53] As translated in: Albert Einstein, *Relativity: The Special and the General Theory*, 1920, Chap IX (Section 9), p. 25ff. Emphasis added.

§ 4. The Nature of Light

Interrelationship of Velocity, Wave Length, Frequency, and Energy

We need to know something about the *lightwave-energy-particle* before we can understand this last question.[54] We need to look at the nature of light to see what Einstein and others have missed about light and its Electromagnetic Spectrum (EMS).

In order to further present my view on the confusion pertaining to light in physics, we will present a short outline of the present state of knowledge on light and its velocity, frequency, wave length, and energy connection. Remember, a short outline by its nature cannot give a full description of the complexity of the matter or a full answer to every element.

Wave-Particle Duality

According to present understanding, electromagnetic radiation is emitted and absorbed in tiny "packets" called photons and exhibits properties of both waves and particles. This phenomenon is referred to as the wave–particle duality. The study of light is known as optics.

Light is radiant energy, sometimes referred to as electromagnetic radiation (EMR). Generally EMR is classified by wavelength into radio, microwave, infrared, visible light, ultraviolet, x-rays and gamma rays. Light is generally spoken of as the visible light seen by the human eye. All electromagnetic radiation is believed to only move at the regular velocity of light in a vacuum, such as in the "vacuum" of outer space. According to wave theory higher frequencies have shorter wavelengths, and lower frequencies have longer wavelengths.

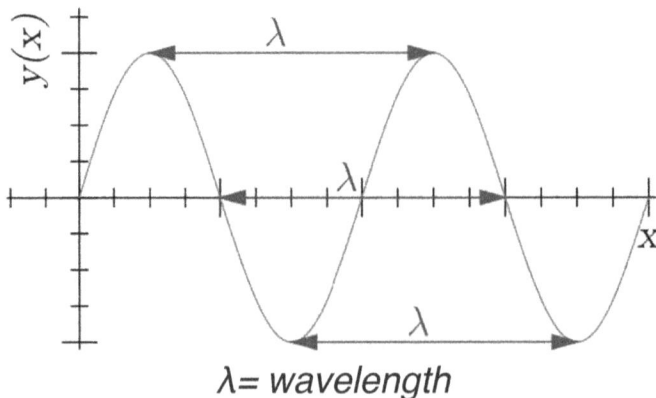

λ= wavelength

Figure 2

---

[54] "Why would M' see the doppler effect instead of the light from B first?'

Light's Wave Formula

$$c = \lambda f$$
velocity = wave length x frequency

Where the Greek lowercase lambda λ equals wavelength, c equals the velocity of light and f equals the frequency of light. The smaller the wavelength the higher the frequency; the longer the wavelength the lower the frequency. Frequency of light is the number of lightwave crests that pass a point every second. Frequency is measured in units of hertz (Hz). 100 hertz = 100 wave crests per second passing a certain point in space. Light travels at about 186,000 miles per second; a 100 hertz would only be 100 crests in 186,000 miles; a 100 megahertz (100 Mhz) would be 100,000,000 crests passing a certain point in one second. If light is thought of as being particles instead of crests (crests of particles), then the frequency of 100 Mhz would be 100,000,000 particles passing a certain point in one second.

Light's Energy, Frequency Relationship

Light's Energy Formula: $E = hf$

E stands for energy in the unite of Joules;
h stands for the unit known as as Planck's Constant;
f stands for frequency of light in units per second (in hertz)

Radio	Microwave Infrared	Visible Light	Ultra-Violet	X-Rays	Gamma Rays	Cosmic Rays

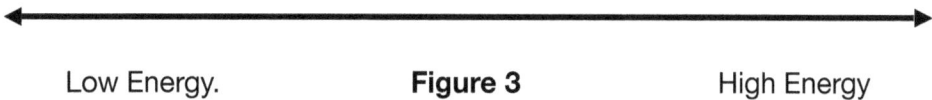

Low Energy.   **Figure 3**   High Energy

Thus, as frequency increases, so does the energy. Frequencies for high frequencies are measured by detectors using indirect methods, that are measuring the *energy* of light — not the frequency of light. **Thus the actual frequency shown on the detector may not correspond to real frequency.**[55]

Joules and Planck's constant are units created for utility in mathematics and science. Other units could also have been used, but the adopted units are useful for communicating ideas and theories because all today's mathematicians use the same units. One thing to remember: if different

---

[55] Frequencies are detected by the energy of light; see under "Instrumental Detection" on the following page.

units[56] or different numerical bases[57] were used then the ratios within equations and formulas would be different and therefore our description of the universe would be different.

Humans' Detection of Light and Color

Humans see light and color when the wave-particles (photons) have enough energy (frequency) to cause a molecular change in the retina of the eye, which triggers an electro-chemical sensation in the optical nerve that is then transmitted to the various visual centers of the brain and then is interpreted by the brain as a color. Higher frequency light such as ultraviolet, X-rays and gamma rays, because of its high energy, is not transmitted to the brain as color. Lower frequency light, because of its low energy, is not interpreted as color or light to the brain and is thus also not registered as light. The cones and rods in the eye only react to certain energy levels (frequencies) of light and transmit these to the brain through the optical nerve. The color of anything is actually the reflection of light wave-particles from an object (which has no innate color itself) to a person's eye, who then interprets the wave-particles as color through a process in the brain. Color is actually only a product of the brain/mind and does not exist in the universe outside of our minds.

Instrumental Detection of Light

Instruments detect light and measure light by detectors[58] in gyros, in frequency counters through the use of quartz, or for very high frequencies the so-called heterodyne mixers. These mixers are used to down shift the light signal from the optical band to an electronically tractable frequency range. These frequency counters and mixers use indirect methods that have more to do with optical photons, **which are usually detected by absorbing the photon's *energy*, thus only revealing the magnitude, and not counting the frequency or measuring the wave lengths**. There are no instruments that are able to count high frequencies or measure extremely small wave lengths. Instrumental detection and measurement of light's frequencies are highly indirect, subjective and complex[59] and rely on the **energy**[60] of the light instead of the equation $(c = \lambda f)$ to ascertain either the

---

[56] System of Units (SI): metric for length, kilogram for mass, second for time, ampere for electric current, kelvin for heat and other base units such as, hertz, newton, joule, watt, volt, ohm, tesla, etc. Another system could have been chosen and thus the relationships of our formulas may change depending on the units we use.

[57] Decimal or binary or sexagesimal or byte or dozenal systems.

[58] These detectors are very complex. See a description of a photodetector used in a Patent filing here: https://encrypted.google.com/patents/US20020089670 [link as of Aug, 2025]

[59] "Optical Heterodyne Detection," accessed Aug. 2025— http://www.rp-photonics.com/optical_heterodyne_detection.html

[60] https://labverra.com/articles/understanding-wavelength-spectrophotometers/ as of August, 2025

wavelength or the frequency. These instruments are calibrated by "known standards."[61] The Hubble Space, James Webb Space, and the Spitzer Space telescopes used such devices as the charge-coupled devices (CCDs) to turn the ultraviolet and infrared light into digital signals.[62]

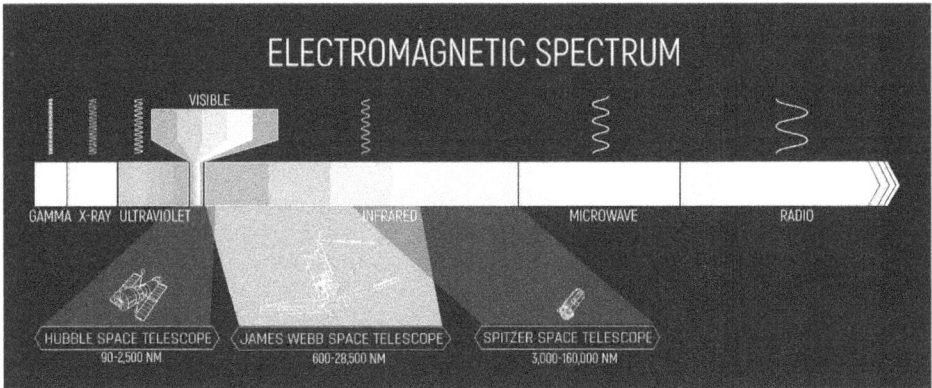

Figure 4: Public Domain graphic[63] on https://webbtelescope.org

Sources of Light

According to theory, light is emitted from a source because its atoms are in an excited state causing the electrons to change their orbit around the nucleus, thus releasing photons of light. The most common light sources are thermal: a body at a given temperature emits a characteristic spectrum of light. As the temperature increases, the peak shifts to higher frequency (shorter wavelengths), producing first a red glow, then a white one, and finally a blue-white color as light moves out of the visible part of the spectrum and into the ultraviolet. Stars and fires are a source of thermal light. Light can also be produced by chemical reaction (fireflies), electroluminescence, thermo-nuclear, etc.

Cosmic Objects & Superluminal Motion

Lasers 300 times faster than the Speed of Light: Also Laser *pulses* can travel 300 times the speed of light.[64]

Cosmic object 14 times the velocity of light: Cosmic objects also have been observed traveling faster than the speed of light. For example, the quasar 3C 279 was reported by Whitney, Shapiro et al. to be expanding "at a

---

[61] https://labverra.com/articles/understanding-wavelength-spectrophotometers/ "Calibration is vital for maintaining the integrity of spectrophotometric measurements. It involves adjusting and verifying the instrument's parameters against known standards."

[62] https://webbtelescope.org/contents/media/images/4188-Image

[63] https://webbtelescope.org/copyright

[64] L. J. Wang et al. 2000 *Nature* **406** 277 — http://physicsworld.com/cws/article/news/2000/jul/19/laser-smashes-light-speed-record

speed of ten times that of light."[65] There were at least 36 "suspected superluminal sources, of which 18 are well established and 10 very probable" as of 1990.[66] The first superluminal motion was observed in 1901. After the Nova Persei's eruption a luminous cloud appeared around the star.

> "For several months it expanded several arc seconds per day. At the nova's distance of 1,400 light-years, the nebula's rate of growth corresponded to a linear speed of 14*c*!"[67] *That is, 14 times the velocity of light.*

> "One way out of the dilemma is to suppose, as do Halton C. Arp and some others, that quasars and related objects are much closer to us than most astronomers believe." [68]

Arp had a different interpretation of the redline shift seen in our heavens.[69] Arp was an American astronomer who worked at the Palomar Observatory for 29 years and at the Max Planck Institute for Astrophysics starting in 1983; he died in 2013.[70]

Great Distances in Space

The only *scientific* method to ascertain the distant to the stars in the universe is by using ground based "trigonometric parallaxes, using the earth's orbit around the sun as a base line, but this method as of 1973 was not accurate beyond about 20 light years. Within our own Galaxy, which is approximately 60,000 light years in diameter, the nearest major galaxy, the Andromeda Nebula, is about 2 million light years away."[71] And The nearest star, Proxima Centauri, was only about 4 light-years from the Sun. With better methods since 1973, recently it is said that up to distance of 300 light years to a star can be ascertained from the ground based telescopes and 10 times further for satellite telescopes. Yet astronomy believes the universe has a radius of about 46 billion light-years. Where do they get this extraordinary belief?

Evolution need vast amount of *time* to work. They found it through the work of Henrietta Swan Leavitt (1868-1921). Her idea to measure

---

[65] Christopher Gregory, "Doppler Effects and Hypervelocities of 3C 279," *Nature Physical Science*, Vol. 239, Sept 25, 1972.

[66] Eric Sheldon, "Faster Than Light?," *Sky & Telescope*, Jan 1990, pp 27.

[67] Ibid., p. 26.

[68] Ibid., p. 27.

[69] George B. Field, Halton Arp & John N. Bahcall, "The Redshift Controversy," 1973, W. A. Benjamin, Inc Advanced Book Program, London. And see — Halton Arp, *Seeing Red: Redshifts, Cosmology and Academic Science*, Apeiron, Montreal, Canada, 1998.

[70] Dennis Overbye, "Halton Arp, 86, Dies; Astronomer Challenged Big Bang Theory," *The New York Times,* Jan 6, 2014, p. B15. http://www.nytimes.com/2014/01/07/science/space/halton-c-arp-astronomer-who-challenged-big-bang-theory-dies-at-86.html

[71] Halton Arp, *The Redshift Controversy*, p. 16, 1973.

astronomical distances led to a shift in understanding the scale of the universe. Her work led to a method of measuring the distance to other stars and galaxies. She discovered a relationship between the luminosity and the period of Cepheid variable pulsates. Although what the brightness of a pulsating star has to do with distance is puzzling, yet astronomers felt it was an answer. Before galaxies were recognized as galaxies, they were believed to be just clouds of gas or dust (Nebula) instead of grand gathering of stars. But Edwin Hubble took Leavitt's work, her "Leavitt's Law," and calculated that the the Cepheids in several nebulae were too far apart to be part of the Milky Way and thus must be separate galaxies. At that time the Milky Way was believed to be the only galaxy.

> "His observations, made in 1924, proved conclusively that these nebulae were much too distant to be part of the Milky Way and were, in fact, entire galaxies outside the Milky Way galaxy; thus, today they are no longer considered nebulae," but galaxies.... Hubble went on to estimate the distances to 24 extra-galactic nebulae, using a variety of methods. In 1929, Hubble examined the relationship between these distances and their radial velocities as determined from their **redshifts**. All of his estimated distances are now known to be too small, by up to a factor of about 7."[72]

Theory of Redshift Phenomenon and Distance to Galaxies

Since about 1929, and because of Edwin Hubble's papers, the redshift in galaxies was believed to indicate distance since a light source moving away from us shifts the light toward the red end of visible light's spectrum. The more the shift to the red end of the optical spectrum the further away the galaxies are, according to the theory.

But as Halton Arp and others have pointed out it could also point to the light losing energy (tired light).[73] The **second law of thermodynamics** says all energy is dissipating, running down, or that heat (energy) flows from hotter (energized) to colder (less energized) regions of matter or space. Light detecting devices measure the *energy* of the light in order to detect its theoretical wave length and frequency. Scientists measure and ascertain the different manifestations of light by its energy. The devices don't and can't detect light's wave lengths or frequencies except with devices that measure it's energy. Furthermore, we don't see the supposed trillions of star images that is believed to be billions of light years away except if we use powerful devices like the Hubble Space or James Webb Space telescopes because

---

[72] https://en.wikipedia.org/wiki/Edwin_Hubble My emphasis; https://pmc.ncbi.nlm.nih.gov/articles/PMC522427/?page=3 links good as of August, 2025

[73] Halton Arp. Seeing Red: *Redshifts, Cosmology and Academic Science*, 1998; Field, Arp, Bahcall, a*The Redshift Controversy*, 1973, W. A. Benjamin, Inc. Advanced Book Program, Reading, Mass. See also: Is the Universe Expanding? Fritz Zwicky and Early Tired-Light Hypotheses, Helge Kragh, Niels Bohr Institute, Blegdamsvej 17, 2100 Copenhagen, Denmark from www.archive.org.

humans can only see about 7000 stars at night on average.[74] If there are trillions of stars, why aren't our night sky totally full of light? It is because the light from stars loose their energy on the way to our earth (Arp's "tired light"[75]). If their energy did not diminish our night sky would be as light or more so than the day sky. A flashlight or searchlight are examples. It's light diminishes over distance equal to it original power/energy.[76] Star light looses it energy[77] because of the second law of thermodynamics, and because of its interaction with the universe's neutron sea. Also, there is no real "space" in the universe. It is only called space because we cannot see the "substances" of "space." Space is full of gravity-fields, photons, neutrons, and probably an invisible "substance" which we call the "Aether-Field."

Light's Velocity Variability

Slower than the velocity of Light: The velocity of light is reduced when traveling through substances such as water to 75% of its 'normal' velocity, 65% in optical fiber or even to almost a complete stop in a cloud of atoms at an almost absolute-zero temperature by a process perfected by a team led by a Harvard professor, Lene Hau.[78,79] Two years later they actually stopped the velocity of light by beaming a laser through a dense cloud of rubidium and helium gas. "The light bounced from atom to atom, gradually slowing down until it stopped. No super-vacuum or ultra-cold was needed."[80] So in a sense the light was transferred to matter (atoms) and later from matter back into light.[81]

A 2015 report of an experiment threw water on the idea that light's velocity in a *vacuum* was a universal constant. By forcing photons through a special mask (a software controlled liquid crystal device), which changed the shape of the forced photon, Daniel Giovannini, et al. showed that one could

---

[74] *Yale Bright Star Catalog,* Both hemispheres about 9100 or maybe 6000-8000 for average eye sight.

[75] See also: Is the Universe Expanding? Fritz Zwicky and Early Tired-Light Hypotheses, Helge Kragh, Niels Bohr Institute, Blegdamsvej 17, 2100 Copenhagen, Denmark from www.archive.org.

[76] see p. 91; https://x.com/i/grok?conversation=1965269278963138834

[77] 2nd law of thermodynamics

[78] William J. Cromie, "Researchers now able to stop, restart light," *Harvard University Gazette,* Jan 24, 2001. http://www.news.harvard.edu/gazette/2001/01.24/01-stoplight.html [Visited Aug, 2025]

[79] Phillip Ball, "Turning light into matter," *Nature,* Feb. 8, 2007. http://www.nature.com/news/2007/070205/full/news070205-8.html [visited Aug. 2025]

[80] Ibid., Cromie

[81] Ginsberg, Garner & Hau, "Coherent control of optical information with matter wave dynamics," *Nature,* Feb 8, 2007, pp. 623-626. — http://www.seas.harvard.edu/haulab/publications/pdf/Ginsberg-Garner-and-Hau-Nature-445-623-(2007).pdf; and see Ibid., Cromie. [Visited Aug. 2025]

reduce light's speed in a *vacuum* permanently even after light exited the mask.[82] Normally, when light's speed is reduced going through a medium (*e.g.* water), it regains its velocity after exiting the the medium. Although the change in velocity in this experiment was very small, it negated the belief that light always moves at its 'normal' velocity in a vacuum.

Does Light's Velocity Vary in a Gyroscope?

Ring lasers are composed of two beams of light of the same polarization traveling in opposite directions in a closed loop. Currently ring lasers are used as gyroscopes in moving vehicles like cars, ships, planes, and missiles. Navigation systems require three gyroscopes to measure pitch, roll, and yaw.

The laser gyro senses rotation and change in direction by sending two laser beams in opposite directions around a path cavity formed by mirrors. When the gyro is at "rest" in a vehicle moving in uniform motion (without acceleration[83]) the laser beams have identical frequencies. When the gyro moves, because the vehicle accelerates, from its initial state, the frequencies are increased in one direction and decreased in the other direction. **The difference between the frequencies of both beams is used to detect the directional movement of the vehicle.[84] This phenomenon is called the *Sagnac effect*.[85] The ring laser measures acceleration and from this a change in direction can be calculated.**

Figure 5

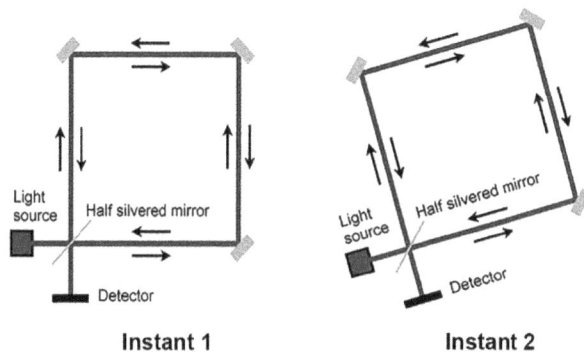

Instant 1   Instant 2

[82] " Spatially structured photons that travel in free space slower than the speed of light," Daniel Giovannini, Jacquiline Romero, Václav Potoček, Gergely Ferenczi, Fiona Speirits, Stephen M. Barnett, Daniele Faccio, and Miles J. Padgett. *Science*. Published online 22 January 2015. [DOI:10.1126/science.aaa3035]

[83] Remember, **acceleration is** the instantaneous rate of change of the velocity and/or direction of a system.

[84] See Dana Z. Anderson, "Optical Gyroscopes," *Scientific American*, April 1986, pp. 94-99.

[85] https://en.wikipedia.org/wiki/Sagnac_effect

The **Figure 6** above on the left is the gyro at **Instant 1**. The graphic on the right above is the same gyro at **Instant 2** after it turned counterclockwise. The light moves at the velocity of light in both directions in **Instant 1**. In **Instant 2** the light going clockwise is moving faster relative to the gyro cavity (has a higher frequency) while the light going counterclockwise is moving slower relative to the gyro cavity (has a lower frequency). The same laser beam moved out from the same light source at the same time. The beam is split by the first mirror ("Half silvered mirror") with half of the beam being reflected to the left (thereafter going clockwise) and half of the beam being passed straight through the same mirror (thereafter going counterclockwise). Each beam moved in opposite directions. At **Instant 1**, since the gyro had not yet moved, both beams moved at the velocity of light (relative to the gyro) and both beams showed the same frequency when received at the detector. But at **Instant 2** the clockwise beam was moving faster than the speed of light relative to the gyro and thus was detected by the detector with a higher frequency than at **Instant 1**. And at **Instant 2** the counterclockwise beam was moving slower than the speed of light relative to the gyro and thus was detected by the detector with a lower frequency than at **Instant 1**. This thus demonstrates that light can and does move faster and slower than the "normal" velocity of light relative to a single observer (the gyro's detector) in the same system (the gyro).

Is the light actually going faster than its normal velocity inside a gyroscope? Yes, *relative* to the cavity of the loop. Is the higher frequency related to its higher velocity? Yes. See next item where we examine an experiment that shows the same phenomenon. Let's examine this more.

Experiment Shows Relativity of Light's Velocity

This is an experiment (and at the same time a thought-experiment) as originally proposed in 1971-72,[86] which manifests that light's velocity can be *relatively* faster than the normal velocity of light.

Figure 6 is a diagram of the purposed experiment. Points **A** & **B** are a stationary system while contraption **A'** & **B'** is a one-piece system with the ability to move along a track in an **A'** to **B'** direction as well as the opposite direction. The distance between **A** and **B** equals the distance between **A'** and **B'**. The light source is enclosed in the box; light is allowed out of this enclosure through two pinholes, **h** and **h'**, at the front of the box. These pinholes are situated so that when light is allowed through them, it will be transmitted to the light sensitive receptors on **A** and **A'**. These receptors create an electric current when light is received on them. This current travels through twowires or two glass cables (from **A** and **A'**) of equal length to a device which measures which receptor received the light first, or if they received the light at the same time.

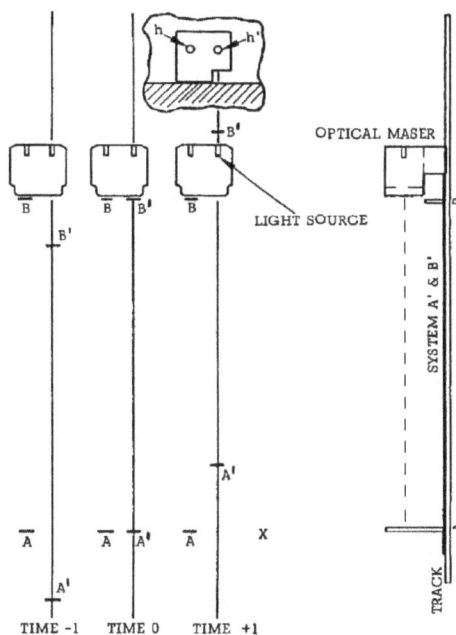

Figure 6

The experiment begins by first testing the timing device. In order to do this, system **A** & **B** and **A'** & **B'** are placed in the position shown in **Figure 6** at time **0**. At this position the pinholes are opened simultaneously and the light travels from its source to **A** and **A'**. If everything is operating properly the timing device should indicate that **A** and **A'** received the light at the same instant. Next system **B'** and **A'** are moved toward the light source as indicated in **Figure 6** at **Time +1**. Again the light is allowed through the pinholes, at the same instant, and thus the timing device should indicate that **A'** received the light first. The same process is repeated with the position shown in **Time -1**. The result from this position should indicate that **A** received the light first.

---

[86] In my first 1971 paper, *The Totality of Motion's Relativity.*

Next the moving system is placed at the extreme end of the track (Fig. 6, Time -1) away from the light source with **A'** facing toward the light source. This system is then "shot" toward the light source at a uniform velocity. When **A'** of the moving system aligns with position **X** (Fig. 10, Time 0) it will set off a current that will allow the light to proceed through pinholes **h** and **h'**, at the same time toward the receptors **A** and **A'**. This tripping device should be systematically set to allow the light to proceed only when systems **A** & **B** and **A'** & **B'** are situated as shown at **Time 0**. Thus light is allowed through the pinholes only when **A** and **A'** are equal distances from the light source.

Between the **Time 0** and **+1** light wave-particles will travel across **B** and **B'** toward **A** and **A'**. If it is possible for light's velocity to be relative, then at **Time +1**, the first light wave-particle should reach **A'** before **A** and this would be thus recorded with the timing device. But because the one-piece system (**A'** & **B'**) is moving toward the light source between **Time 0** and **+1**, the light thus travels across the known distance of the one-piece contraption (**B'** & **A'**) faster than between the same known distance **B** and **A**. The relative velocity between distance **B'** And **A'** would be found to be the velocity of light *plus* the velocity of the moving system.

But notice, please, that although the light would travel from **B'** to **A'** faster than the normal velocity of light, it would be observed at **A'** at **Time +1** at light's normal velocity for although it travels across the moving one-piece contraption faster than the velocity of light, it travels the distance from the pin holes to the receptors at exactly the normal velocity of light. If there was an observer on **A'**, he could calculate that the light has traveled from the light source to him in its normal time and velocity. We too can perceive this. Yet we also can see that in this experiment light travels from **B'** to **A'** relatively faster than from **B** to **A** even though the distance between **B'** and **A'** and **B** and **A** are exactly the same.

Doppler Effect

This experiment could, would and does prove the relativity of light's velocity. **Although at any instant A' could calculate the light[87] at the normal speed of light, the light would have the Doppler effect and would be shifted toward the blue light**, thus manifesting that its relative velocity is faster than the normal speed of light. This phenomenon is the Sagnac effect that manifests itself in laser gyroscopes (see § 4).

---

[87] Light with all its colors (frequencies).

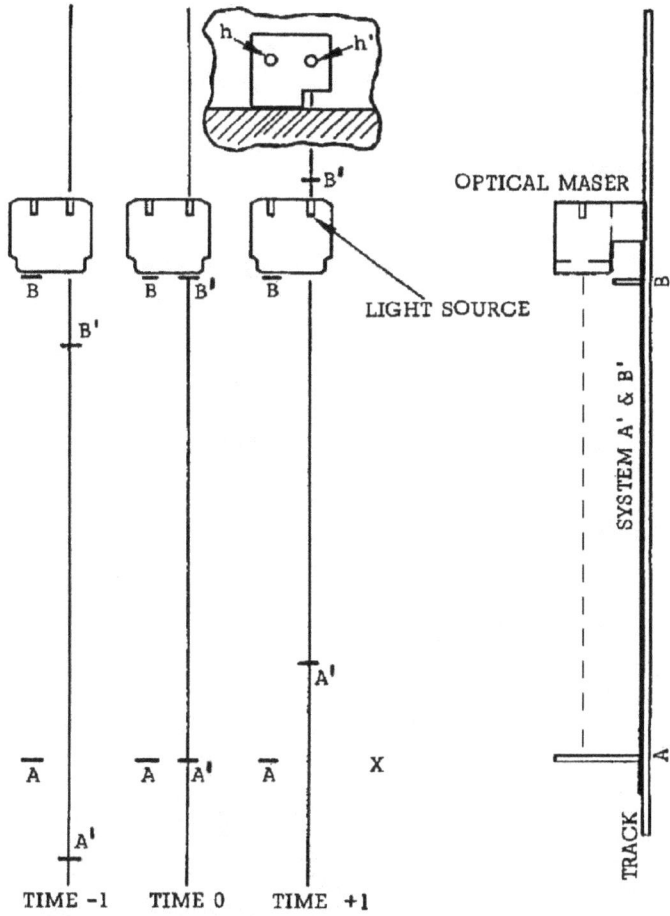

Figure 6

Yes, Light's velocity is Independent of source

Light's velocity is independent of source. According to the second premise of special relativity: "light is always propagated in empty space with a definite velocity c which is independent of the state of motion of the emitting body."[88] Or as Daniel Frost Comstock put it in 1910, "The second postulate is that the velocity of light is independent of the relative velocity of the source of light and observer."[89] The evidence used to support this postulate concerns the behavior of binary stars.

Over the years various articles have been printed on the astronomical evidence of this postulate.[90] Arguments against this evidence have been presented by H. Dingle and others.[91] F.G. Fox has pointed out the importance of the extinction theorem of dispersion theory.[92] According to T. Alväger et al.:

> This implies that, as the radiation from the moving source enters a stationary refractive medium, it is progressively absorbed and replaced by similar radiation re-emitted by the medium. As the medium is stationary the new radiation must have a normal velocity, and all effects of the motion of the source (if any) will be lost.[93]

Furthermore, according to T. Alväger et al., because of this, binary stars:

> cannot be regarded as conclusive tests of the second postulate of Special Relativity. The same can be said about other astronomical observations.

> Partly because of this, much effort has recently [1960s] been devoted to terrestrial tests of the second postulate. Many of these remain, however, either of low accuracy or are subject to the same criticism as the astronomical observations.[94]

---

[88] From Einstein's Premise 2.

[89] *Science,* May 20, 1910 **31**:768. Found online @ https://archive.org/details/science45sciegoog. Comstock is an American physicist.

[90] W. De Sitter, *Physik. Z.* **14**, 429, 1267 (1913); E. Freundlich, *Physik. Z.* **14**, 835 (1913); P. Gutnick, *Astron. Nachr.* **195**, 265 (1913); W. Zurhellen, *Astron. Nachr.* **198**, 1 (1914); W. De Sitter, *Bull. Astron. Inst. Netherlands* **2**, 121 (1924); M. LaRosa, *Z. Physik* **21**, 333 (1924), and *Z. Physik* **34**, 698 (1925); W. E. Bernheimer, *Z. Physik* **36**, 302 (1926); H. Thirring, *Z. Physik* **31**, 133 (1925).

[91] H. Dingle, *Mon. Nat. R. A. S.* **119**, 67 (1959); P. Moon and D. E. Spencer, *Opt. Sec. Am.* **43**, 635 (1953).

[92] J. G. Fox, *Am. J. Phys.* **30**, 297 (1962); *Am. J. Phys.* **33**, 1 (1965).

[93] T. Alväger et al., *Arkiv Fysik* **31**, 145 (1966).

[94] T. Alväger et al., *Arkiv Fysik* **31**, 145 (1966).

T. Alväger et al. go on to explain their experiment and its result, which they indicate conforms to the second postulate of relativity and disproves the emission theory (that the velocity of light is dependent on its source).

Also see and read other explanation of Einstein's Special Theory of Relativity in wikipedial.org

In this paper we will assume there is enough evidence that light's velocity is independent of its emitting source.

What does it mean that light's velocity is independent of its emitting body? Simply, it means that the velocity of the emitting body is not added to or subtracted from the velocity of the emitted light. When each light wave-particle leaves its emitting body it travels at the apparent speed of light.

Doppler Effect and Relative Velocity

Now here is another important point that I wish to make: The principle of the second postulate as worded by Einstein, and quoted above, with qualifications[95] is true. But the velocity of light can be independent of its emitting body *and* also received by an observer/device relatively slower or faster by another body moving toward or away from the emitting body system. This means in certain cases that the velocity of light is not constant to all observers. This slower or faster relative velocity is manifested by the doppler effect. Let me amplify.

---

[95] Depending on our perception and knowledge of light itself.

§ 5. Confusion Between the Nature of Light and Velocity

Figure 7 shows the Doppler effect in two opposite directions. On the left the waves are of a lower frequency (further apart from each other) than normal because the source is moving away. The right side of the graphic shows the waves at a higher frequency because the radiating source is moving in the right direction and each wave is catching up with the one in front.

Figure 7

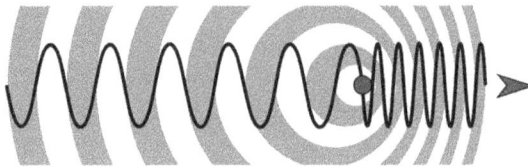

In Figure 8 below, a star is moving toward a relatively motionless planet.[96] The common distances between the light wave-particles, when released by the star, would look like those in instants 1, 2 and 3. The light received by the observer on the planet, will have the doppler effect because the star is moving toward the planet. The photons each instant is released closer to the one released the previous instant because of the star's movement toward the planet, thus giving the doppler effect of a higher frequency/energy. The velocity of *each* light-wave particle is its normal velocity when released, but because of the relative movement of the star they seem bluer to the observe on the planet. If the velocity of the star was great, the observer on the planet may not see any light because the doppler effect would manifest the light outside the visual spectrum.

Figure 8

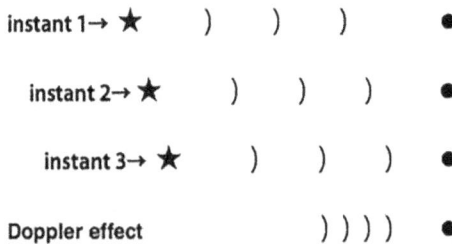

Study the following Figure 9 and its sequent paragraphs

---

[96] Remember in a multi-bodied universe you can figure relative motion. See § 1.

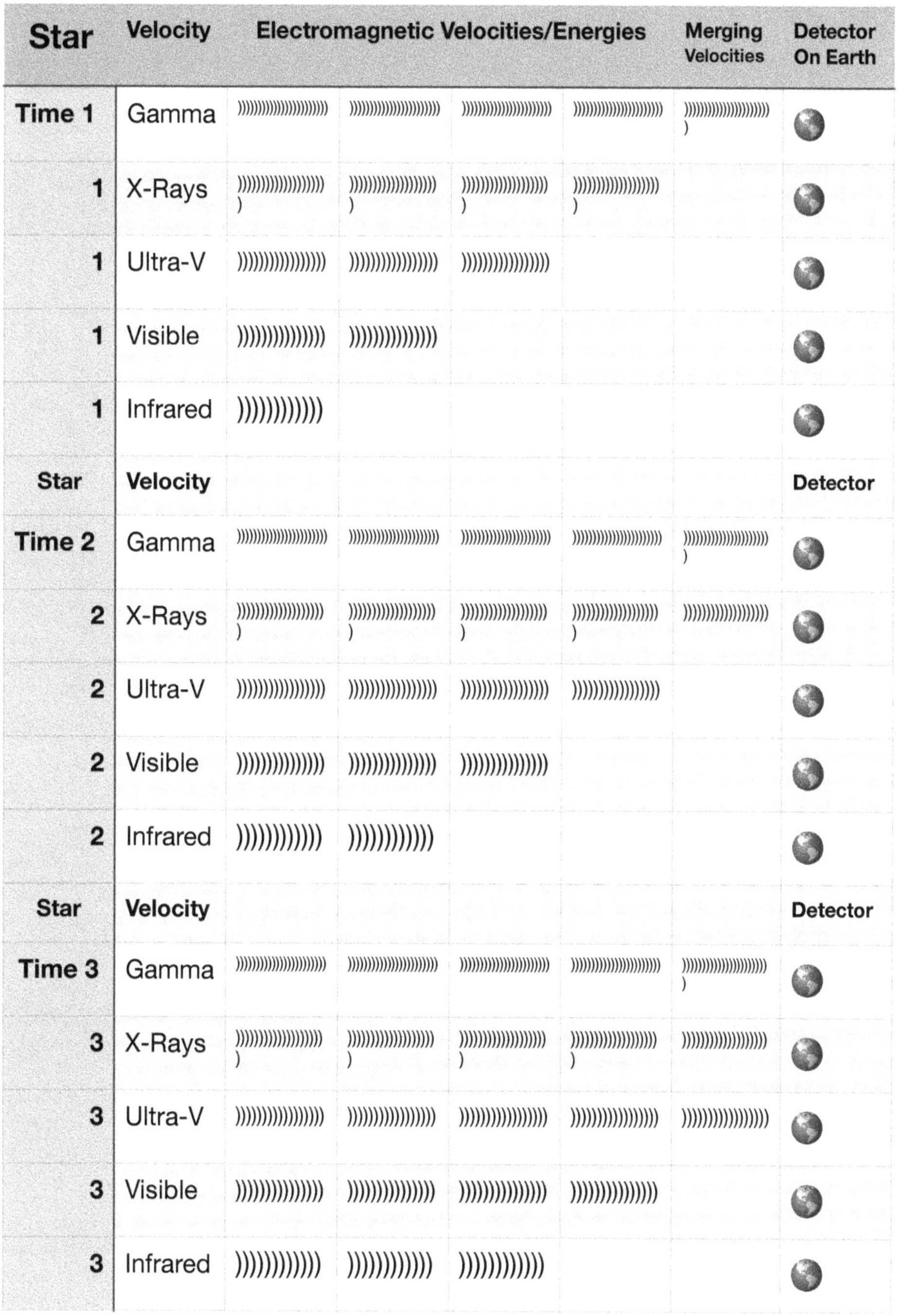

Star	Velocity	Electromagnetic Velocities/Energies				Merging Velocities	Detector On Earth
Time 1	Gamma	)))))))))))))	)))))))))))))	)))))))))))))	)))))))))))))	)))))))))))))	🌎
1	X-Rays	)))))))))))))	))))))))))))	)))))))))))))	)))))))))))))		🌎
1	Ultra-V	))))))))))))	)))))))))))	)))))))))))			🌎
1	Visible	))))))))))))	))))))))))))				🌎
1	Infrared	))))))))))					🌎
Star	Velocity						Detector
Time 2	Gamma	)))))))))))))	)))))))))))))	)))))))))))))	)))))))))))))	)))))))))))))	🌎
2	X-Rays	)))))))))))))	)))))))))))))	)))))))))))))	)))))))))))))	))))))))))))	🌎
2	Ultra-V	))))))))))))	)))))))))))	)))))))))))	)))))))))))		🌎
2	Visible	))))))))))))	))))))))))))	))))))))))))			🌎
2	Infrared	))))))))))	))))))))))				🌎
Star	Velocity						Detector
Time 3	Gamma	)))))))))))))	)))))))))))))	)))))))))))))	)))))))))))))	)))))))))))))	🌎
3	X-Rays	)))))))))))))	)))))))))))))	)))))))))))))	)))))))))))))	))))))))))))	🌎
3	Ultra-V	))))))))))))	)))))))))))	)))))))))))	)))))))))))	)))))))))))	🌎
3	Visible	))))))))))))	))))))))))))	))))))))))))	))))))))))))		🌎
3	Infrared	))))))))))	))))))))))	))))))))))			🌎

Figure 9: Graphic by Walter R. Dolen

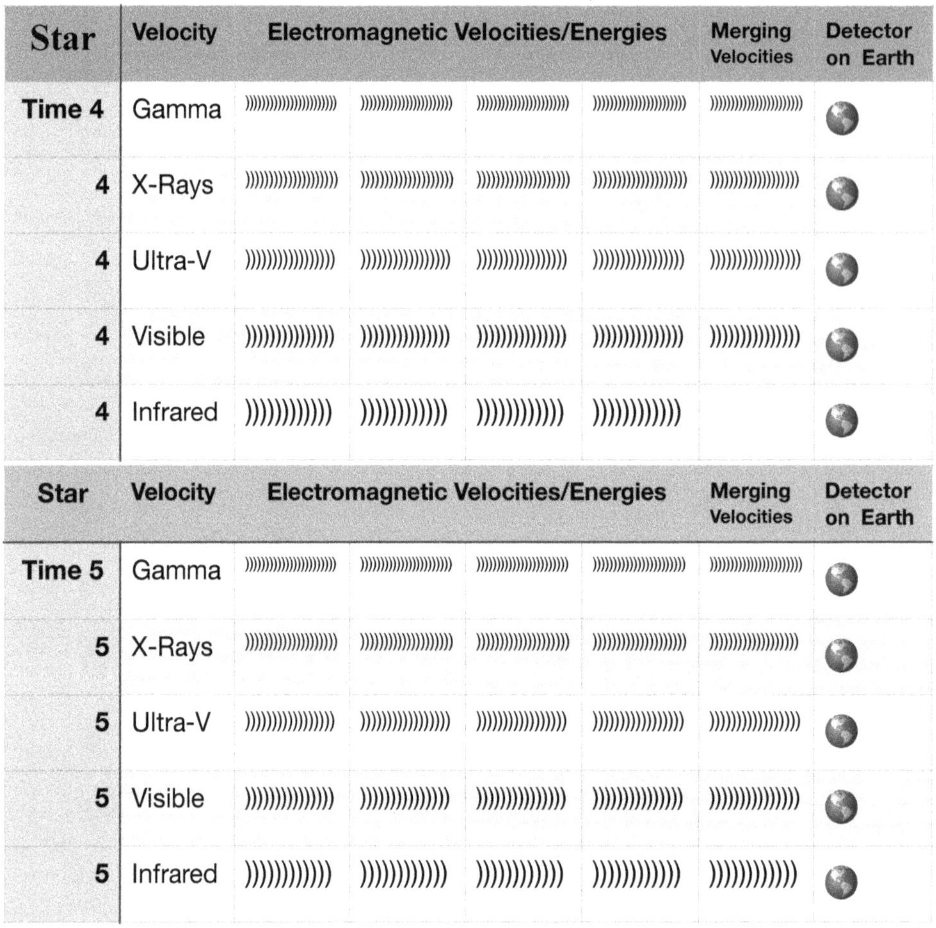

Star	Velocity	Electromagnetic Velocities/Energies				Merging Velocities	Detector on Earth
Time 4	Gamma	)))))))))))))	)))))))))))))	)))))))))))))	)))))))))))))	)))))))))))))	🌎
4	X-Rays	)))))))))))))	)))))))))))))	)))))))))))))	)))))))))))))	)))))))))))))	🌎
4	Ultra-V	)))))))))))))	)))))))))))))	)))))))))))))	)))))))))))))	)))))))))))))	🌎
4	Visible	)))))))))))))	)))))))))))))	)))))))))))))	)))))))))))))	)))))))))))))	🌎
4	Infrared	)))))))))))))	)))))))))))))	)))))))))))))	)))))))))))))		🌎

Star	Velocity	Electromagnetic Velocities/Energies				Merging Velocities	Detector on Earth
Time 5	Gamma	)))))))))))))	)))))))))))))	)))))))))))))	)))))))))))))	)))))))))))))	🌎
5	X-Rays	)))))))))))))	)))))))))))))	)))))))))))))	)))))))))))))	)))))))))))))	🌎
5	Ultra-V	)))))))))))))	)))))))))))))	)))))))))))))	)))))))))))))	)))))))))))))	🌎
5	Visible	)))))))))))))	)))))))))))))	)))))))))))))	)))))))))))))	)))))))))))))	🌎
5	Infrared	)))))))))))))	)))))))))))))	)))))))))))))	)))))))))))))	)))))))))))))	🌎

Figure 9: Graphic by Walter R. Dolen

Electromagnetic Velocities/Energies appear to merge

In Figure 9, and probably in reality, Electromagnetic Velocities/Energies appear to merge like the colors within visible light seem to merge as the color white. As we see at **Time 1**, the more energized rays from the Star reach the earth first. Next at **Time 2, 3 and 4**, the less energized rays begin to catch up. As time goes by all the different energized rays will **seem** to be reaching the earth at the same time as shown in **Time 5**. But in reality, for each distance between the star and earth, the more energized rays reach the earth earlier. This phenomenon can be compared to the mixing of the visible light colors that reaches the Earth as white light, which is revealed to us by the Prism as mix of different colors (wave lengths) or energies.

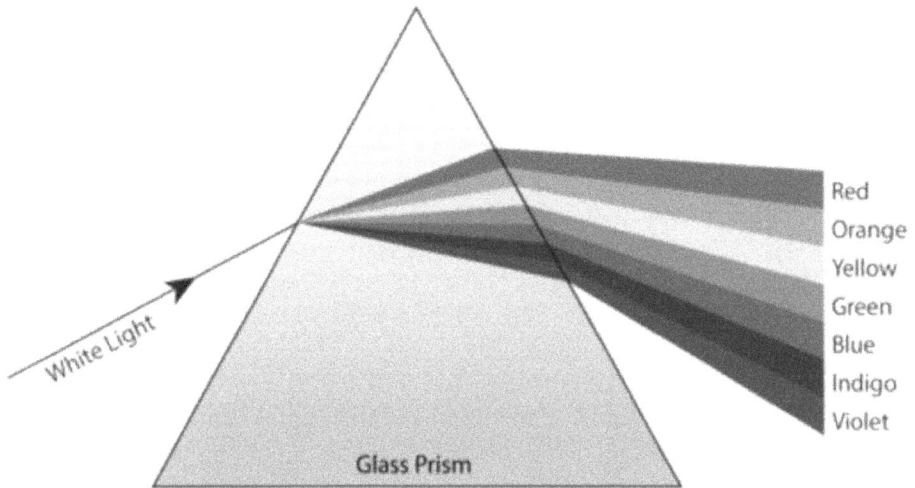

Figure 10: Image by Jibin 1840404

Creative Commons Attribution Share Alike 4.0

The more energetic color (violet) has more energy and velocity than the slower Red color. If we had a prism for all the "colors" of light from Gamma Rays to Radio "waves," we would understand better that the energy of Electromagnetic Spectrum translates to the various velocities for each energy level, or "wave-length."

Remember the sophisticated light detectors devices all measure **energy**, NOT wave length or frequency. There is no way of proving that our idea of frequencies and wavelengths are the reality because we have no way of measuring them, just like we still don't have any way of proving our ideas of electrons, or the nucleus, or the atom, or that atoms are all the same size.[97]

Experiment to Ascertain the Various Velocities of Light

Full Eclipse of the Sun. Have 5 observer devices pointed at the moon at least seven minutes before the eclipse ends and the sun's rays begin to appear from the back of the moon. Each device can only read **one** of the electromagnetic frequencies: Gramma Rays, X-Rays, Ultra Violet Rays, Visible light, or Infrared Rays. The observer devices must be connected to a detector that identifies which EMS rays from the sun reaches the devices first to last, and in what order, as soon as the first Gamma Rays from the sun peek around the moon. Due to the atmosphere's absorption of the EMS the best results would come from satellite devices located above the earth's atmosphere.

More information on this section § 5 will be found in the near future at https://beone.ws/light=energy.

----

[97] see Appendix 7

§ 6. An Aether-Field in the Universe

Einstein replaced the idea of an aether with the idea of "field." Aether, or luminiferous aether, was a theoretical substance proposed in the 19th century as a medium for the propagation of light and its electromagnetic waves. The concept was abandoned after such experiments as the Michelson-Morley experiment (1887) failed to detect it. To many the development of Einstein's theory of relativity provided a better explanation for the behavior of light without the need for aether.

Einstein:

> More careful reflection teaches us, however, that the special theory of relativity does not compel us to deny ether. We may assume the existence of an ether; only we must give up ascribing a definite state of motion to it....[98]

In other words, he didn't think that such experiments as Michelson-Morley's disproved the idea of aether, but only that the experiment didn't prove there was anyway to prove absolute motion.

My examination of the Michelson-Morley experiment in my 1971-1972 papers showed that the gravity-atmosphere[99] may have been the medium for the light in the experiment, not the aether, and that is why it didn't indicate absolute movement. In other words, light from the experiment was affected by either gravity or the atmosphere of the Earth or both, not the speculative aether surrounding the earth.

Acceleration and Inertial

For Einstein the "field" was needed to answer the phenomenon of acceleration and Inertia.

Einstein knew one could use the entire universe as a reference system. In 1917 he wrote:

> The most important fact that we draw from experience as to the distribution of matter is that the relative velocities of the stars are very small as compared with the velocity of light. So I think that for the present we may base our reason upon the following approximative assumption. **There is a system of reference** relatively to which matter may be looked upon as being permanently at rest. [100]

---

[98] Albert Einstein, *Sidelights on Relativity,* an address delivered on May 5th, 1920, in the University of Leyden. Amazon sells a copy of this on their website as of 2014.

[99] light from space was/is (1) affected by the sun's gravity (1919), (2) is affected by aberration from star light, (3) is affected by water refraction.

[100] A. Einstein, "Cosmological Considerations on the General Theory of Relativity," IX, § 3, 1917 found in — A. Einstein, *The Principle of Relativity: A Collection of Original Papers on the Special and General Theory of Relativity*, 1952, Dover edition. Emphasis added.

In context, Einstein was using the whole universe as a reference system. Einstein later in his life in 1950 said the same thing when writing about acceleration — which needs to be referenced to something:

> **"Inertia resists acceleration, but acceleration relative to what?** Within the frame of classical mechanics the only answer is: Inertia resists acceleration *relative to space.*"[101]

But Einstein in this paper seems to say that acceleration in his General Theory is accelerated relative to the "continuous field" represented in his theory by the **coordinates of space and time**:

> "According to general relativity, the concept of space detached from any physical content does not exist. The physical reality of space is represented by a **field** whose components are continuous functions of four independent variables — coordinates of space and time."

In order to answer the problem of inertial and acceleration, we will substitute the words, **Aether-Field,** for the invisible "substance" of Space and not use , **aether**, or **coordinates of space and time** or **field**.

Neither, "aether" or "field" or "space" or "coordinates of space" will work for me because of their past usage in physics. So what is the "aether-field."

Aether-Field

- (1) Aether-Field always existed.

  Since there is no scientific or philosophical way to conceptualize a *beginning* to the Aether-Field it must have been existing always:[102]

- (2) Energy is all the motion within the Aether-Field.

  Since there is no such thing as solids or particles of solids,[103] then there is only energy moving in orbits of different sizes, orbits within orbits, ("wheels within wheels") that manifest as apparent solids within our eyes or our devises. We cannot see the Aether-Field, only the energy within it. That is why the Aether-Field is invisible.

- (3) The "something else" to explain inertia/acceleration must be the Aether-Field itself and the other forms of energy manifesting within the Aether-Field.

  There can be no motion except when compared to the "something else"

- (4) Aether-Field is somehow highly intelligent.

---

[101] A. Einstein, "On the Generalized Theory," *Scientific American*, April 1950 [reprint], p. 5. Emphasis is in text.

[102] read reprint of Grok's analysis— Appendix 8

[103] Particle-wave phenomenon: particles of what; waves in what?

Because of the sophistication and complexity of energy forms within the Aether-Field, and their knowledge-based codes (DNA, RNA, etc.), there must be high intelligence within the Aether Field.

Therefore:

- (5) Everything (the energy) in the universe is from, within, and exists because of the Aether-Field's intelligence.[104]

Conclusions From Our Study of Einstein's Special Theory of Relativity (STR)

(1) We live in a multi-bodied universe; there is no such thing as a stationary object. All objects are in relative motion.

(2) The equation — $E = mc^2$ — is unsubstantiated and did not originate with Einstein. (§ 1 and Appendix 1 of this book)

(3) Both Premises of Einstein's Special Theory of Relativity (STR) misrepresented reality; we restate them as:

   (A) Premise 1: Although today there is no available method of detecting absolute motion, because this is a multi-bodied universe, we can ascertain the relative motion of all bodies in the universe by using the totality of the energy in the Aether-Field, the phenomenon of the inertia-acceleration effect, and Galilean transformations.

   (B) Premise 2: Light's velocity is independent of its source and can be *relativity* faster or slower than its "normal" velocity, to certain observers: the Doppler Effect manifests this relative velocity.

(4) The theory's calculations vary from the classical system of relativity *only* when the velocity of a body is near the speed of light as defined by STR.

(5) The theory, because of it usage of Lorentz's transformation formula, leads to major contradictions in logic and meaning of words (time, length) which are impossible to the rational mind or to a realistic universe.

---

[104] see Aether-Field

(6) The theory's idea of time is amiss from the historical notion of time. Time is in reality a comparative quality of the mind (as is color[105]) that is non-relative by definition and tradition; the examples given by the adherents of the theory did not prove the relativity of simultaneous events, but manifest their spatial confusion of the matter and time and light.

(7) The General Theory of Relativity with its four dimensional template, the theory of black holes in astronomy and the idea of antimatter are all *ad hoc* explanations to explain away problems generated because of the problematic underpinning of Einstein's Special Theory of Relativity.[106]

(8) Light may move relatively faster and slower than its normal velocity[107] and the Doppler effect is one indication of this. When looking directly at a *relative* superluminal velocity, we would also not see it as observable light, since the Doppler effect would turn it into a higher frequency outside of the visual spectrum.[108] Light moving relatively slower than the normal velocity of light would also not be seen as observable light. It would be out of the visual spectrum at the lower end (microwave, infrared).

(9) If light can move faster than its normal velocity,[109] and if the light's superluminal velocity is directed at us (not laterally[110]), our eyes will *not* be able to see it because the energy-frequency would be too great for our eyes to detect it as light. Our scientific instruments (which only measure energy) would also not be able to detect it as moving faster than light. We might, for example, indirectly observe this superluminal phenomenon as X-rays or gamma rays or cosmic rays instead of visible light.

---

[105] Color does not exist in the universe outside of our mind; it is a product of a process in our eyes and brain.

[106] See Appendixes 3 & 4.

[107] As shown in § 4, 5 & 6.

[108] See Figure 7, 8 & 9 and their surrounding text and footnotes.

[109] Possible because a light source may generate light at a much higher energy level than our sun. Remember frequency and energy are related.

[110] When you look at light laterally, such as a flashlight pointed at a right angle to your eyes, you are looking at the reflection off of the molecules of the gas or air, not directly at the stream of light.

(10) **From the information in this paper, one can reasonably conclude:**

- The velocity of light is not a universal constant.
- There is nothing motionless or stationary in the universe
- All motion is relative — the totality of motion's relativity.
- Other theories derived from the Special Theory of Relativity (STR), such as the General Theory (with its four dimensions), the black holes of modern astronomy, quantum's antimatter and other offsprings of STR are not necessarily real. This virtual universe may only live in the minds of mathematicians[111] because of the weak foundation of STR.
- **Aether-Field**, as defined above,[112] sounds a lot like the God mentioned in the Bible who is everywhere, all powerful and all knowledgable, and is the God, all in all: the was, the is, and the will-be One; "nothing besides me, I am the Yehova and there is nothing else." (Isa 45:6).

---

[111] See the Appendices in this book.

[112] Aether-Field, p. 46-48, (5)

§ 7. Appendices — Assumptions of Science

1. Relativity of Mass?
2. Contradiction of Clocks
3. Fourth Dimension
4. General Theory, Black Holes & Antimatter
5. GPS and Relativity
6. Mathematics and Reality
7. Science is a Guessing Game
   - Heisenberg and Dirac tell their stories about the guessing game in their own words, from their 1968 lectures, as published in Abdus Salam's book

   - Basic Assumptions of Quantum Theory are Arbitrary
   - Scientists Observe Through Their Theories

   - Constancy in Science Was Often Only Assumptions

8. Why is There Anything
   - Recent Theories (as of 2025) on the Universe's Origin
   - Flashlight's/Searchlight's Diminishing Power: Energy = Distance
9. Original papers on Einstein's relativity theories

Appendix 1: Relativity of Mass?

Pertaining to the theory's idea that mass will increase as it moves closer to the speed of light, we note a few things. Let Barnett explain this aspect of Einstein's theory:

> Einstein's equation giving the increase of mass with velocity is similar in form to the other equations of Relativity but vastly more important in its consequences:

$$m = \frac{m_0}{\sqrt{1 - \frac{v^2}{c^2}}}$$

> Here m, stands for the mass of a body moving with velocity v, m_0 for its mass when at rest, and c for the velocity of light. Anyone who has ever studied elementary algebra can readily see that if v is small, as are all the velocities of ordinary experience, then the difference between m_0 and m is practically zero. But when v approaches the value of c then the increase of mass becomes very great, reaching infinity when the velocity of the moving body reaches the velocity of light. Since a body of infinite mass would offer infinite resistance to motion the conclusion is once again reached that no material body can travel with the speed of light.

> Of all aspects of Relativity the principle of increase of mass **has been most often verified and most fruitfully applied by experimental physicists**. Electrons moving in powerful electrical fields and beta particles ejected from the nuclei of radioactive substances attain velocities ranging up to 99 per cent that of light. **For atomic physicists concerned with these great speeds, the increase of mass predicted by Relativity is no arguable theory but an empirical fact their calculations cannot ignore**.[113]

Notice that he says "the principle of increase of mass has been often verified and most fruitfully applied by experimental physicists." Scientists usually mention experiments inside particle accelerators to support their belief. Is this true?

Accelerators increase the velocity of charged particles by creating large electric fields which attract or repel the particles. This field is then moved down the accelerator, "pushing" and "pulling" the particles along. Modern accelerators, such as the Stanford Linear Accelerator in California can generate up to 50 giga electron-volts of energy (50 GeV) to accelerate the

---

[113] Lincoln Barnett, *The Universe and Dr. Einstein*, 2nd Revised Ed., 1957, pp. 56-57. Emphasis added.

particles.[114] In so doing it is adding energy, thus mass, according to Einstein's theory. It is not the velocity that increases the mass, but the energy-mass that is put into the accelerator to accelerate the particles. Therefore use of mass-accelerators do not support the theory, yet this is the very thing they use to prove their idea. In addition, as we see below, modern physicists no longer believe in the popular equation (E = mc²) used in textbooks or popular articles.

Infinite mass or no mass? Here is where those backing the theory use another obvious *ad hoc* explanation to save their theory. In the famous equation, E = mc², mass (*m*) times the velocity of light (*c*) squared equals the total energy (*E*) of that body. So one would think logically that when a photon of light is going at the speed of light that its mass would be almost infinite. Yet when a photon from the Sun hits the Earth it does not strike with the force it would if the equation were true. What do the Einsteinists do to save their theory? They make up an *ad hoc* statement: "light [photons] has no mass," yet at the same time say it has energy, pressure and momentum — just not the energy of infinite mass. Light as an electromagnetic wave, although it carries no mass, does carry energy. It also has momentum, and does exert pressure (radiation pressure). This is the reason tails of comets point away from the Sun. The main reason scientists believe that photons (light) do not have mass is that if they did, according to Einstein's equation, they could not move at the exact speed of light in vacuum because they would have almost infinite mass; every light-particle in the universe cannot have infinite mass.

Physicists have two definitions for mass: one is "relativistic mass" and the other "invariant mass." They call the relativistic mass the old definition. It gives every object a velocity-dependent mass: m = E / c². Now, since that would make no sense, they use the new definition where every object just has one mass, an invariant quantity that does not depend on velocity: m= E_0 / c², where E_0 is the total energy of that object at rest. The first (old) definition is often used in popularizations, and in some elementary textbooks. Most physicists use the new definition. The new definition is now called the "rest mass," or the "invariant mass." The "relativistic mass" is never used at all. "It is almost certainly impossible to do any experiment which would establish the photon rest mass is exactly zero."[115]

---

[114] The center-of-mass is 100GeV divided by 2 = 50 GeV. See "prospects for High Energy..." by R.B. Palmer, Stanford Linear Accelerator Center, March 1990, p 4 — found on Stanford Linear Accelerator Center's web site: http://www.slac.stanford.edu/pubs/slacpubs/5000/slac-pub-5195.pdf (as accessed in Dec 2014).

[115] Matt Austern et al., "What is the mass of a photon?," (Usenet Physics FAQ) as accessed Jan. 2015 — http://math.ucr.edu/home/baez/physics/ParticleAndNuclear/photon_mass.html

"In the modern language of relativity theory there is only one mass, the Newtonian mass m, which does not vary with velocity; hence the famous formula $E = mc^2$ has to be taken with a large grain of salt."[116]

$E = mc^2$ is actually $E_0 = mc^2$. In a 2009 paper by L. B. Okun of the Institute for Theoretical and Experimental Physics, Moscow:

> The concept of relativistic mass, which increases with velocity, is not compatible with the standard language of relativity theory and impedes the understanding and learning of the theory by beginners. The same difficulty occurs with the term rest mass. To get rid of relativistic mass and rest mass it is appropriate to replace the equation $E = mc^2$ by the true Einstein's equation
>
> $E_0 = mc^2$, where E_0 is the rest energy and m is the mass.[117]
>
> Unfortunately, sometimes and especially in his popular writings Einstein was careless about the subscript $_0$ and spoke about the equivalence of mass and energy and omitted the attribute "rest" for the energy. As a result Einstein's equation **$E_0 = mc^2$** became known in its famous but misleading form **$E = mc^2$**. One of the most unfortunate consequences is the concept that the mass of a relativistic body increases with its velocity.[118]

Clarification of the Equations Symbols. Einstein used $E_0 = mc^2$ (Einstein used different letters in his equations than modern usage) in one of his other 1905 papers, "Does the inertia of a body depend on its energy content?," as cited by Okun. But Einstein's original paper on relativity did not use $E_0 = mc^2$, but $E = mc^2$ (using different letters in the equation) when figuring the kinetic energy of the electron.

---

[116] Lev B. Okun, "The Concept of Mass," *Physics Today*, June 1989, p. 31. Lev Okun was head of the laboratory of elementary-particle theory at the Institute of Theoretical and Experimental Physics, in Moscow in 1989. PDF copy of paper can be found at (accessed on Jan 2015): www.itep.ru/theor/persons/lab180/okun/em_3.pdf

[117] L. B. Okun, *Am. J. Phys.*, Vol. 77, NO. 5, May 2009, p. 430.

[118] Ibid., p. 431.

Appendix 2: Contradiction of Clocks

Herbert Dingle, a Professor of History and Philosophy of Science at the University of London and a former president of the Royal Astronomical Society (1951-53),[119] after believing in Special Theory of Relativity for 40 years, had an awaking in the early 1960s. He studied relativity for about 40 years, learning it from the late Professor A. N. Whitehead and writing a book, *Relativity for All*, in 1921. He had discussions with all those physicists whose names are best connected with Special Theory of Relativity, such as Einstein, Eddington, Tolman, Whittaker, Schrödinger, Born, Bridgman, et al. Yet Dingle came to the realization that the theory was absolutely false. He tried for over 10 years to get his fellow professors and scientists to rethink their position with little success. Therefore in 1972, after having many of his arguments published in *Nature*, he finally wrote a book, *Science at the Crossroads*. Most scientists did not understand his position and thus rejected it.

The contradiction of clocks was pointed out by Herbert Dingle and misunderstood by many.[120] Because no comments, pro or con, were received about his twice-published criticism, Dingle later repeated his criticism in *Nature* and asked the readers to respond. He showed in this paper that:

> "The argument used to prove that 'moving clocks run slow' (with which all the kinematical implications of the theory are bound up) proves, with exactly the same validity, that moving clocks run fast. Both cannot be right, so the basis of the theory must be faulty."[121]

One reply to Dingle's criticism was printed in this same issue of Nature, by Max Born.[122] According to J. G. Fox,[123] the general conclusion of Born's reply to Dingle represents "the majority opinion in physics." If this is true, it means the majority of physicists do not understand Einstein's theory, for Born did not understand Dingle's criticism.

Dingle, in his criticism published by *Nature*, quotes the passage from Einstein's 1905 paper which purports to prove that moving clocks run slow. With this passage Dingle writes a parallel passage which proves moving clocks run fast.

In Born's reply to Dingle, he concludes:

---

[119] Dingle was one of the founders of the British Society for the History of Science, and served as President from 1955 to 1957. He founded what later became the British Society for the Philosophy of Science as well as its journal, the *British Journal for The Philosophy of Science*.

[120] H. Dingle, *Phil. Sci.* **27**, 233 (1960); Samuel, Viscout, and H. Dingle, *A Threefold Cord* (Allen and Unwin, 1961), p. 270.

[121] H. Dingle, *Nature* **195**, 985 (1962).

[122] Max Born, *Nature* **197**, 1287 (1963).

[123] J. G. Fox, *Am J. Phys.* **33**, 16 (1965).

"The two cases are therefore different, the symbols t and τ in expressions (1) and (2) referring to different physical situations; these are inverse and must of course correspond to an exchange of the symbols t, and τ. This is exactly what expressions (1) and (2) express. There is no contradiction."[124]

Born therefore believed Dingle's two expressions represented different physical situations, and this therefore accounted for the contradiction.

But in Dingle's reply to Born, printed right after Born's letter, he shows where Born is confused:

What Einstein's statement shows is that, on his theory, between the two events, (1) the coincidence of the clocks, and (2) a later event occurring on X, the time-interval according to A is greater than that according to X, so that X is running slower than A. What my statement shows is that, in the identical physical situation, between the two events (1) the coincidence of the clocks, and (2) a later event occurring on A, the time-interval according to A is smaller than that according to X, so that X is running faster than A. The only difference between the two cases is that the intervals compared are those between different pairs of events; and the only conceivable way of avoiding the contradiction is by showing (consistently with the relativity postulate, that is, without assuming an aether in which A is uniquely stationary and X uniquely moving) that Einstein's interval between two events on X is valid for comparing the rates of the clocks, and that my interval between two events on A is not.[125]

Notice, carefully, in Dingle's reply what he has in parentheses: "without assuming an aether in which A is uniquely stationary and X uniquely moving." The reason that Born believes the two equations of Dingle represent different physical situations, is because he thinks that A is uniquely stationary and X is uniquely moving. But to think this is to disclaim the first premise of Einstein's theory — that no motion is absolute because there is apparently no absolute at rest reference system (*i.e.* aether). Since according to the first premise of special relativity, no one can ascertain if A is uniquely stationary or not, then Dingle's criticism is correct, and Born's is wrong.

Hence Dingle showed, that in Einstein's original paper, Einstein had no logical reason in his example to arbitrarily say that one coordinate system was "stationary" and the other "moving." Because of the first postulate of special relativity, three choices are possible:

(1) K is "stationary" and k is moving;

(2) K is moving and k is "stationary;"

---

124 Ibid., Born.

125 H. Dingle, *Nature*, **197**, 1288 (1963).

or (3) $\underline{K}$ is moving and $\underline{k}$ is moving.[126]

Thus Dingle asked in a previous paper:

> *why, consistently with the theory* [first postulate], *the former result* [from possibility (1)] *must be accepted as true, while the latter* [from possibility (2)] *must be rejected as false.*[127]

In accordance with the first premise of special relativity (ignoring the fact that we live in a multi-bodied universe[128]), one of the two clocks in Einstein's example in his original paper moves at once faster and slower than the other clock. Why? Because Einstein just arbitrarily called one system stationary: "we call it the 'stationary system.'" He could have called the other reference system stationary, or, if one cannot ascertain which reference system (the train or embankment) was moving, or if both were moving, he could have arbitrarily said one system had this or that velocity while the other had that or this relative velocity related to the other. Therefore, from one example one could envision an infinite number of clock times, according to the theory. As Einstein wrote when explaining the "relativity of simultaneity," in his book, *Relativity: The Special and General*:

> Every reference-body (co-ordinate system) has its own particular time; unless we are told the reference-body to which the statement of time refers, there is no meaning in a statement of the time of an event.[129]

This is utter nonsense. There would be an infinite number of "local times." This would mean there is no universal time. Without universal time there would be no way of measuring a single velocity of light. The velocity of light would be different for every reference system in the universe because each system would use different lengths and times (local times). Just as the geocentricists, with their mathematical system, hung on to their faulty system, the Einsteinists are hanging onto their illogical system. It is time for scientists to move on. The present generation of scientists apparently cannot move on because they have too much at stake — their status, their reputations. Yet this is what science is — a continuum of theories hopefully moving to a better understanding of the universe. For science to be science, it must move on beyond the weak and contrary underpinning of Einstein's relativity theories.

---

[126] There is nothing stationary in the universe; everything is in relative motion to everything else. By knowing we live in a multi-bodied universe we can approximately ascertain the relative motion of everything if we have the right information and technology. See § 1.

[127] H. Dingle, *Nature*, **195**, 986 (1962). Emphasis in text.

[128] Einstein ignored it in his example.

[129] A. Einstein, "IX The Relativity of Simultaneity," in *Relativity: The Special and General Theory*, 1920, p 26.

Local clocks are only a consequence or a possibility because of the Special Theory of Relativity. If the foundation of the theory (constant velocity of light for all observers) is incorrect, the theory fails.

Appendix 3: Fourth Dimension

Hans Reichenbach, a philosophy of science educator, who in 1920 became an instructor in physics, and eventually associate professor, at the Technische Hochschule in Stuttgart, and who later taught physics at the Berlin University between 1926 and 1933, and who also taught philosophy at the University of Istanbul and later at the University of California in Los Angeles, wrote the following in his book, *The Philosophy of Space & Time*:

> Whereas the conception of space and time as a four-dimensional manifold has been very fruitful for mathematical physics, its effect in the field of epistemology has been only to confuse the issue. **Calling time the fourth dimension gives it an air of mystery**. One might think that time can now be conceived as a kind of space and try in vain to add visually a fourth dimension to the three dimensions of space. It is essential to guard against such a misunderstanding of mathematical concepts. If we add time to space as a fourth dimension, it does not lose in any way its peculiar character as time. **Through the combination of space and time into a four-dimensional manifold we merely express the fact that it takes four numbers to determine a world event**, namely three numbers for the spatial location and one for time. [...] **Our schematization of time as a fourth dimension therefore does not imply any changes in the conception of time**.[130]

Time in Einstein's theories is merely a mathematical way of looking at space when taking into consideration the time aspect. There is no universe with four or more dimensions. The fourth dimension was merely a mathematical invention to explain the universe IF the velocity of light is constant and never relative. This paper indicates otherwise. Light's velocity can be *relatively* faster or slower than the "normal" velocity. And light's velocity itself (outside of the idea of *relative* velocity) probably can be faster or slower than the "normal" velocity except it may be detected as X-rays/gamma rays/cosmic rays or radio waves as explained in § 4 & 5. Time is a quality generated by processes in our brains. We see and organize the world around us through the concept of time. Time is the "software" used by us to perceive the universe in a chronological manner.

---

[130] Hans Reichenbach, *The Philosophy of Space & Time*, Translated by Maria Reichenbach and John Freund, Dover, 1958, pp. 110-111, originally written in 1927-28. Emphasis added.

Appendix 4: General Theory, Black Holes & Antimatter

Special Relativity is the basis for the General Theory of Relativity, the theory of black holes and the theory of antimatter. If Special Relativity is wrong, it is likely those beliefs based on it are wrong too.

General Theory

Einstein once stated that "the development of the theory of relativity proceeded in two steps, 'special theory of relativity' and 'general theory of relativity.' The latter presumes the validity of the former as a limiting case and is its consistent continuation."[131]

Since the formulas of the general theory of relativity are based on the special theory of relativity, then the general theory is wrong if the special theory is incorrect, or there is a great likelihood of this. I believe there is enough evidence herein to disprove Einstein's special theory.

Black Holes

The famous theoretical physicist Kip S. Thorne wrote:

> All the properties of the black hole are determined **completely** by Einstein's Laws for the structure of empty space.[132]

"Thorne's research has focused on gravitation physics and astrophysics, with emphasis on relativistic stars, black holes and gravitational waves. In the late 1960's [sic] and early 70's [sic] he laid the foundations for the theory of pulsations of relativistic stars and the gravitational waves they emit. During the 70's [sic] and 80's [sic] he developed mathematical formalism by which astrophysicists analyze the generation of gravitational waves and worked closely with Vladimir Braginsky, Ronald Drever and Rainer Weiss on developing new technical ideas and plans for gravitational wave detection."[133]

---

[131] Albert Einstein, *Out of My Later Years* (New York: Philosophical Library, Inc., 1950), p. 42.

[132] Kip S. Thorne, "The Search for Black Holes." *Scientific American*, Dec 1, 1974, p.35. Emphasis added.

[133] "Kip Thorne received his B.S. degree from Caltech in 1962 and his Ph.D. from Princeton University in 1965. After two years of postdoctoral study, Thorne returned to Caltech as an Associate professor in 1967, was promoted to Professor of Theoretical Physics in 1970, became The William R. Kenan, Jr., Professor in 1981, and The Feynman Professor of Theoretical Physics in 1991." Bio taken from the Caltech web page — http://www.its.caltech.edu/~kip/scripts/biosketch.html (as accessed in Dec. 2014).

Antimatter

We will quote from Abdus Salam's[134] book *Unification of Fundamental Forces*:

> May I now turn to the problem of the elementary particles. I think that really the most decisive discovery in connection with the properties or the nature of elementary particles was the **discovery of antimatter by Dirac**. That was **an entirely new feature which apparently had to do with relativity,** with the replacement of the Galilei group by the Lorentz group. **I believe that this discovery of particles and antiparticles by Dirac has changed our whole outlook on atomic physics completely.** I do not know whether this change was realized at once at that time, probably it has been accepted only gradually; but I would like to explain why I consider it so fundamental.
>
> We know from quantum theory that, for instance, a hydrogen molecule may consist of two hydrogen atoms or of one positive hydrogen ion and one negative hydrogen ion. Generally one can say that every state consists virtually of all possible configurations by which you can realize the same kind of symmetry. Now as soon as one knows that one can create pairs according to Dirac's theory, then one has to consider an elementary particle as a compound system; because virtually it could be this particle plus a pair or this particle plus two pairs and so on, and **so all of a sudden the whole idea of an elementary particle has changed.** Up to that time I think every physicist had thought of the elementary particles along the line of the philosophy of Democritus, namely by considering these elementary particles are unchangeable units which are just given in nature and are just always the same thing, they never change, they never can be transmuted into anything else. They are not dynamical systems, they just exist in themselves.
>
> **After Dirac's discovery everything looked different, because now one could ask, why should a proton be only a proton, why should a proton not sometimes be a proton plus a pair of one electron and one positron and so on.**[135] [pp. 114-115]

The General Theory of Relativity with its four dimensional template, the theory of black holes in astronomy and the idea of antimatter are all *ad hoc* explanations to explain away problems generated because of the problematic underpinning of Einstein's Special Theory of Relativity. If the Special Theory of Relativity is incorrect, the theories based on it are more than likely also faulty.

---

[134] Abdus Salam's work in theoretical physics has been crucial to the development of modern particle physics. His pioneering work was in the renormalisation theory for mesons, the two-component theory of the neutrino, dispersion relations, unitary symmetry, electro-nuclear unification, proton decay, supersymmetry and superspace. In 1979, he shared the Nobel Prize for Physics with Steven Weinberg and Sheldon Glashow for work on the unification of the electromagnetic and weak interactions.

[135] *Unification of Fundamental Forces: The First of the 1988 Dirac Memorial Lectures* by Abdus Salam [Cambridge University Press, Cambridge, GB], 1990, pp. 114-115. Emphasis added.

Appendix 5: GPS and Relativity

Note: the following is a quote from a 1996 paper by Fliegel and DiEsposti, "GPS and Relativity," which shows how little Einstein's equations differ from the classical transformation equations and thus are negligible to proper operation of the Global Positioning System. GPS systems do not need to be adjusted by Einstein's equations to work properly. As we have shown in this book, only when the velocity of an object approaches the speed of light is there any real difference between Einstein's relativity and classical relativity.

GPS AND RELATIVITY: AN ENGINEERING OVERVIEW

Henry F. Fliegel and Raymond S. DiEsposti

The Operational Control System (OCS) of the Global Positioning System (GPS) does not include the rigorous transformations between coordinate systems that Einstein's general theory of relativity would seem to require - transformations to and from the individual space vehicles (SVs), the Monitor Stations (MSs), and the users on the surface of the rotating earth, and the geocentric Earth Centered Inertial System (ECI) in which the SVs orbits are calculated. There is a very good reason for the omission: **the effects of relativity, where they are different from the effects predicted by classical mechanics and electromagnetic theory, are too small to matter less than one centimeter, for users on or near the earth.** [p. 189] ...

In this paper, we compare the predictions of relativity to those of intuitive, classical, Newtonian physics; we show how large or small the differences are, and how and for what applications those difference[s] are large enough to make it necessary to correct the formulas of classical physics. [p. 189] ...

As we have shown, introducing the y factor makes a change of only 2 or 3 millimeters to the classical result. In short, there are no "missing relativity terms."[136] [p.197]

---

[136] Henry F. Fliegel and Raymond S. DiEsposti, "GPS and Relativity: An Engineering Overview," GPS Joint Program Office, The Aerospace Corporation. Paper online at: http://tycho.usno.navy.mil/ptti/1996papers/Vol%2028_16.pdf (as accessed in Dec. 2014). Emphasis added.

Appendix 6: Mathematics and Reality

There is flexibility in mathematics to do almost anything. For example, Ptolemy's geocentric system was a mathematical theory that viewed the Earth as the center of the solar system, if not the universe. This system was used for over a 1000 years to predict the future position of planets and to predict eclipses. But it was fundamentally wrong. Yet this completely false theory's mathematics worked to bring much precision to astronomy. Just as the language of words can be used to write bizarre fictional stories, mathematics is a language of numbers in which you can do many things with numbers that are not necessarily real. Despite what some mathematicians expound: mathematics is only a language of quantities — **not**, as Max Tegmark said in his 2014 book:

> The universe ultimately is mathematics. I'm going to argue that our physical world not only is *described* by mathematics, but that it is mathematics: a mathematical structure, to be precise."[137]

This is total nonsense!

Mathematics is only a language and not as Einstein believed:

> One reason why mathematics enjoys special esteem, above all other sciences, is that its laws are absolutely certain and indisputable, while those of all other sciences are to some extent debatable and in constant danger of being overthrown by newly discovered facts.[138]

Einstein sounds like Ptolemy in his *Almagest*, where Ptolemy wrote:

> Only mathematics can provide sure and unshakeable knowledge of its devotees, provided one approaches it rigorously. For its kind of proof proceeds by indisputable methods, namely arithmetic and geometry.[139]

Remember, Ptolemy was a proud mathematician and his geocentric system was called the "mathematical systematic treatise" in the Greek title of his work. It was believed by the most learned men for over a 1000 years. "It was dominant to an extent and for a length of time which is unsurpassed by any scientific work except Euclid's *Elements*."[140] It was not a religious concept as popularly expounded today. As some scientists do today, Ptolemy cheated and fudged some of his figures. He found a pattern and filled in the

---

[137] *Our Mathematical Universe*, p. 6.

[138] *Sidelights on Relativity* (1922), pp. 25-56, London: Methuen.

[139] *Ptolemy's Almagest*, translated and annotated by G.J. Toomer, Springer-Verlag, NY, 1984, p. 38.

[140] Ibid., as noted by Toomer in the Preface, p. 2.

missing numbers without observation,[141] as did some of the scientists in the genesis of the new physics (see Appendix 7: "Science is a Guessing Game").

Yes, to the mathematicians their craft is the alpha and omega of truth, but the facts are that their math is full of arbitrary manipulations, *ad hoc* rules to save their previous equations and theories, and uses imaginary and negative numbers unrelated to reality. In their mindset they have become blind by their virtual reality world. They are no better than the religiosity they are so fond of disparaging. We need to move beyond it as we have moved beyond Ptolemy's mathematical system, which was also supposed to be the perfect system because it was based on the "certain and indisputable" world of mathematics.

Math is a Language

Richard Feynman, an eminent theoretical physicist, known for his work on quantum mechanics, quantum electrodynamics, particle physics and so forth, said this about mathematics: "Mathematics is just a language."... "Mathematics is a language plus reasoning; it is like a language plus logic. Mathematics is a tool for reasoning." ... "Mathematics is just organized reasoning."[142]

As German mathematician R.L.E. Schwarzenberger said:

> My own attitude, which I share with many of my colleagues, is simply that mathematics is a language. Like English, or Latin, or Chinese, there are certain concepts for which mathematics is particularly well suited: it would be as foolish to attempt to write a love poem in the language of mathematics as to prove the Fundamental Theorem of Algebra using the English language.[143]

Yes, mathematics is only a language. Some aspects of our universe can be better communicated with words, other aspects by the use of numbers. As words can be misused, so can math and numbers. The language of words is factual if it describes real events, real things. If not it is fictional. The language of math is factual and scientific if its numbers describe real things. If not, it is fictional.

Arithmetic and Real Things

Mathematics has to do with manipulation of numbers. Numerals are the symbols that humans use in oral or written communication to represent numbers and quantity. Numbers are used for measuring, counting, ordering, for representing quantity and ratios of quantities. The most basic use of numbers is for counting items such as apples: 1 for 🍎; 2 for 🍎🍎; 3 for 🍎🍎🍎

---

[141] Robert R. Newton, *Ancient Planetary Observations and the Validity of Ephemeris Time,* The John Hopkins University Press, 1976, pp. 147ff.

[142] Richard Feynman, *The Character of Physical Law*, The M.I.T. Press, 1965, 1990 printing, pp. 40-41. He won the Nobel Prize in physics in 1965.

[143] R. L. E. Schwarzenberger, *The Language of Geometry, A Mathematical Spectrum Miscellany*, Applied Probability Trust, 2000, p. 112.

and so on. A basic form of mathematics is arithmetic. Arithmetic deals with numbers, specifically whole numbers 1, 2, 3 and so on. In arithmetic numbers can be used for addition, subtraction, multiplication and division. Numbers can be manipulated by addition. 1 apple 🍎 plus 2 apples 🍎🍎 equals 3 apples 🍎🍎🍎. Or thus, 1 + 2 =3. Or numbers can be manipulated by subtraction: 3 apples 🍎🍎🍎 minus 2 apples 🍎🍎 equals 1 apple 🍎. Thus, 3 — 2=1.

But notice this: 1 triangle ▲ plus 2 stars ★★ equals 3 items of mixed content: ▲★★, not 3 triangles ▲▲▲ or 3 stars ★★★. Notice that numbers, by themselves, only represent quantity. We normally use numbers to represent real things and real items, not imaginary things. Numbers are used by us to quantify and/or proportionate (*e.g.* ⅓ or ½ of something) what we see in the universe. And like time and color,[144] numbers are only something within our minds. Numbers do not exist in the universe. If math does not represent real things, it cannot be used to correctly describe the real universe. Such non-reality based math can only describe a virtual universe, not the real universe. If numbers and math do represent real things, they are then only an aid to perceive the universe — not a part of the universe itself. Unlike the Pythagoreans of ancient Greece, who believed that the world was, quite literally, generated by numbers, in reality numbers and mathematics only exist in the mind and are used by us to quantify and proportionate the world around us.

Arbitrary Rules in Math

Rules of mathematics can be and often are arbitrary. For example in multiplication the product of two positive or whole numbers is positive for two apples times two apples equals 4 apples. Thus the rule, "If both numbers are positive, their product is a positive number." But it makes **no** sense that minus two apples times minus two apples equals plus four apples, which is what you get with the rule: "the product of two negative numbers is positive." Nor does it make sense that minus two apples times plus two apples equal minus four apples, which is what you get with the rule, "the product of two numbers of opposite sign is negative."

Math Rules

Where do mathematical rules come from? From a universal cosmic law that penetrates all matter as many mathematicians seem to believe? Who makes these rules? Have these rules been scientifically proven? Are there exceptions? Are there contradictions in the rules? Are they arbitrary?

Alberto A. Martínez said the following in his book, *Negative Math*:[145]

Most of us are comfortable in the conviction, with Mr. Smith, that

---

[144] See § 3 & 4.

[145] Alberto A. Martínez, *Negative Math: How Mathematical Rules Can Be Positively Bent,* Princeton University Press, 2006, p. 1.

$$2 + 2 = 4.$$

And some of us might be sufficiently unconcerned with math to be amused by people who, by contrast, wonder about two plus two making three, or five, or all numbers at once....

But what about some other mathematical propositions? Is it really true that
$$-4 \times -4 = 16?$$

What does this mean physically?

Wouldn't it be more logical to say,
$$-4 \times -4 = -16?$$

Where did this math rule come from? The answer is that we do not know. With this rule and others a whole subsystem of mathematics started with the illogical $-4 \times -4 = $ a plus number, 16. What are the consequences of such an illogical rule? This isn't the only illogical rule in mathematics. For instance take, for example, given a box with five apples in it, you cannot physically subtract or remove more than five apples and leave a negative quantity of apples in the box. Yet mathematically you can do this. "Likewise, there is no physical experiment, no measuring device, that when made to obtain a numerical measurement, under whatever circumstance, will yield an imaginary number." [Ibid., Martínez, p. 2]

Imaginary Numbers
An imaginary number is the square root of -1 or, as written in mathematics, $\sqrt{-1}$ or with the symbol *i*. More on this later.

"Many students of elementary math are puzzled initially by irrational, imaginary, and even negative numbers. But teachers usually convince them to accept such numbers as just as valid as ordinary numbers. Students are taught to not be misled by the names of such numbers into supposing that they are in any way less real or logical than ordinary numbers."[Ibid., Martínez, p. 2]

Some, "believe, for all practical purposes, that the elementary foundations of mathematics were discovered long ago, and established so securely that today we need not bother to question the validity of the simplest operations. They presume that if anything remained ambiguous or problematic in the elements of mathematics, then it would have been solved already by some clever fellow back in the nineteenth century."[Ibid., Martínez, p.7]

Question: Why is $-4 + -4 = -8$, but $-4 \times -4 = +16$? Because someone made up this rule years ago and everyone of authority has followed it without questioning the significance of the contradiction.

There is an asymmetry in "standard" algebra, biased against negative numbers. Martínez shows how an alternative algebra can be invented, changing the "laws" of maths to develop a completely new system, just as the mid-1800s saw the invention of new *non-Euclidean* geometries. For example, you can rewrite the standard rule to demand that the product of two negatives is itself a negative. Thus the uneasy problem of imaginary numbers has been simply wiped out — the root of -9 is just -3. Martínez tests other consequences of the made-up rules, leading us on an exploration of his new algebra. This isn't just an academic exercise though; a different system devised by Hamilton, called *quaternions*, evolved into the vector algebra that is now used extensively by physicists to describe the real world in another way.[146]

What Martínez did in his book, *Negative Math*, is to first of all show the arbitrariness of mathematical rules and then he showed that you can actually change these rules to make sense, but with different results. So what is the true math: the one that makes sense, or the one with contrary, illogical rules? Can mathematics be trusted to help us ascertain the true nature of our universe?

Multiple answers for the same equation

The equation, $x^3 = 8$, has three solutions with the only right answer being the principal root **2**, and not $-1 + \sqrt{-3}$, or $-1 - \sqrt{-3}$. How do we know which answer is right? By experience, by observation. Yet if we just use mathematics without experience, how would we decide? Flip a coin? Guess?

"According to how we add or subtract the imaginary terms we then have *four* solutions to the addition of complex numbers. And any further operation with complex numbers will multiply the possible number of solutions accordingly [....] The multiplicity of solutions of some equations is a common feature in various branches of mathematics. For exmple, inverse trigonometric functions have infinitely many solutions."[147]

"To advoid multiple solutions, mathematicians and scientists have employed various conventions,"[148] such as artificaly distinguishing the "principlal root" from other root answers. "Another way to avoid multiple solutions involves the concept of function."[149] The problem is that the mathematicians amongst themselves cannot agree as to the meaning of some of the rules. You would need to be a professional mathematician to know what goes on behind the scenes, as you would have to be a historian to know how the historians disagree among themselves.

As we have noted previously, math is nothing but a language of numbers. Just as you can misuse the language of words, so too with the language of numbers. Let's look at the misuse of both languages.

---

146 Lewis Dartnell, Review of Negative Math, maths.org: http://plus.maths.org/content/negative-math May 31, 2006 as accessed on Oct 27, 2014.

147 Ibid., Alberto A. Martínez, p. 117 & p. 119.

148 Ibid., Alberto A. Martínez, p. 120.

149 Ibid., Alberto A. Martínez, p. 120.

Misusage of the language of words:

No man ever steps in the same river twice?

Heraclitus of Ephesus said a man could not step into the same river twice, since it was always changing:

> *Socrates:*
>
> Heracleitus says, you know, that all things move and nothing remains still, and he likens the universe to the current of a river, saying that you cannot step twice into the same stream.[150]

What Heracleitus says is true, that everything in the universe is in a system of change, but there is a popularized misusage of this quote that is as mistaken as Zeno's misusage of math in our example below. What is the misusage of the language of words? It is this. You might not step into the same water molecules twice, but you will step into the same river twice. A river is the totality of the water flowing in a water-basin valley, not individual water molecules flowing in the basin. There is a slight of hand in the popularized misusage of Socrates' quote. When speaking of the river one is not refering to the molecules in the river, but the totality of the flowing water in the basin.

Misusage of the language of numbers:

Zeno: Motion Impossible?

Zeno of Citium reasoned that motion was impossible, according to Aristotle's Physics (book VI), for a moving object will never reach any given point because however near it may get to the point, it always must first accomplish a halfway stage, and then another halfway stage of what is left and so on. Half of half goes on forever and the object will never reach the goal point. But this is using math in a mistaken way and also a misuse of words to describe the situation. The true way of looking at the problem is that the object is moving *toward* the goal by 1/10, 2/10, 3/10, 4/10, 5/10 until it reaches the goal (10/10). It's a mathematical addition process, not a subtraction process of ½ of ½ of ½. In reality there is motion and a moving object will reach the goal point one additional step at a time. The wrong mathematics was used by Zeno. Zeno used subtraction on a problem that needed addition to be solved. Zeno's misuse of math is also evident in his math problem of the foot race between the tortoise and the great warrior. Achilles, according to Zeno, never catches the slower tortoise. In reality the warrior will pass the slow tortoise. Zeno was a verbal magican who used a mental sleight of hand. Einstein also used a perceptional slight of hand when he tried to discount the reality or possibility of simultanity.[151]

---

[150] Plato, *Cratylus* [402a].

[151] See § 3. In Einstein's case I don't believe he did it on purpose; he actually believed in what he was saying. Zeno, on the other hand, was being an intellectual smart aleck.

How is the field of mathematics a logical and *exact* science, if there are multiple answers to a single equation, if you have imaginary numbers, if you think there is a square root of -1? The theories of the new physics use imaginary numbers and negative numbers. These false concepts of mathematics are the dirt in the engine of science. Yes mathematicians can make some of their dubious math appear to work only by adding *ad hoc* rules and arbitarily picking answers to equations that have multiple answers. They simply guess or manipulate the numbers to keep their theory alive. Most people outside of the various specialized fields in physics and astronomy do not understand their special mathematical language, and simply do not care as long as it does not interfere with their lives. I care because some people actually think there are four or more dimensions and think that contradictions in logic are fine, all of which leads to more illogical and incomprehensable nonsense, of which we have too much in our world.

Appendix 7: Science is a Guessing Game

In many ways science is a guessing game used to find answers about our Universe. A hypothesis is another name for a guess. One should not put too much weight on the advertised precision of science over other disciplines, especially when one examines the inexactitude of some scientific theories.

Guessing in Mathematics by Famous Scientists

The eminent Richard Feynman said this about the famous theoretical physicist Paul Dirac:

> "Dirac discovered the correct laws of relativity quantum mechanics simply by guessing the equation. The method of guessing the equation seems to be a pretty effective way of guessing new laws."[152]

Much of science in academia[153] is not what most of the general public thinks it is. A sizable proportion of it is not exact, not perfect, not based on direct observation, but rather, based on indirect observation, arbitrary and specious mathematics, circular thinking, *ad hoc* adjustments to imperfect theories and just plain wild guesses. There are a few books that can give you an idea about what some of today's science is about, especially in the new physics, astronomy and evolution. One such book is Abdus Salam's *Unification of Fundamental Forces: The First of the 1988 Dirac Memorial Lectures*. This book not only contained Salam's 1988 lecture on the "Unification of fundamental forces," but more importantly to us it contained a printed version of the 1968 lectures given by Werner Heisenberg and Paul Dirac.

Abdus Salam's work in theoretical physics has been crucial to the development of modern particle physics.[154] His pioneering work was in the renormalisation theory for mesons, the two-component theory of the neutrino, dispersion relations, unitary symmetry, electro-nuclear unification, proton decay, supersymmetry and superspace. In 1979, he shared the Nobel Prize for Physics with Steven Weinberg and Sheldon Glashow for work on the unification of the electromagnetic and weak interactions.

Werner Karl Heisenberg was a German theoretical physicist, one of the key pioneers of quantum mechanics. He published his work in 1925 with a breakthrough paper and a series of papers with Max Born and Pascual Jordan in the same year. He published his uncertainty principle in 1927. He was awarded the Nobel Prize in Physics for 1932 "for the creation of

---

[152] Richard Feynman, *The Character of Physical Law*, The M.I.T. Press, Cambridge, Mass. 1967, p. 57.

[153] We say "academia" because the scientists that work for businesses have to come up with ideas that are provable and workable. Not so in academia.

[154] The Official Web Site of the Nobel Prize. As accessed on Feb 26, 2015: http://www.nobelprize.org/nobel_prizes/physics/laureates/1979/salam-bio.html

quantum mechanics."[155] He also made contributions to the theories of hydrodynamics of turbulent flows, the atomic nucleus, cosmic rays, subatomic particles, etc.

Paul Adrien Maurice Dirac was an English theoretical physicist who has "laid the foundations, first of the Theory of Quantum Mechanics, second of the Quantum Theory of Fields, and third — with his famous equation of the electron — of the Theory of Elementary Particles."[156] Dirac shared the Nobel Prize in Physics for 1933 with Erwin Schrödinger for the discovery of new productive forms of atomic theory. He lived 1902-1984.

I'll let Heisenberg and Dirac tell their stories about the guessing game in their own words, from their 1968 lectures, as published in Abdus Salam's book: [157]

Note: The words of the text have the light gray background. All bold lettering in the gray background is my emphasis of their words.[158] My comments below the quotes do not have a gray background and are in a different font.

---

[155] http://www.nobelprize.org/nobel_prizes/physics/laureates/1932/ as accessed on February 2015.

[156] Ibid., Salam, p. 2.

[157] Abdus Salam, *Unification of Fundamental Forces: The First of the 1988 Dirac Memorial Lectures*, Cambridge University Press, Cambridge, GB, 1990.

[158] My book is a critique of science so I will quote their exact words instead of paraphrasing them to make it as clear as possible — the author.

Following Taken from Heisenberg's 1968 Lecture

Phenomenological Theory

"We had a nice group of young people doing phenomenological physics together, that is, inventing formulae which seem to reproduce the experiments." [From Heisenberg's lecture in Salam's book, p. 89]

"Sommerfeld asked me again to translate these results into the language of quantum theory, and it turned out that this was easily done." [From Heisenberg's lecture in Salam's book, p. 90]

Author's comment: What is said here is that the mathematician/scientist invents a formula to reflect their view of the experiments. But, if the experiment is faulty, or the premises of the experiment are faulty, then the formula is faulty.

———

"Bohr[159] had just at that time published his papers on the periodic table of elements and **we learned the very complicated structures of all the elements with ten or twenty or thirty electrons being in different orbits and we could not understand how Bohr could have obtained these results**. We felt that he must have been an infinitely clever mathematician to solve such horrible problems of classical astronomy. We knew that even the problem of the three bodies had not been solved by the best astronomers, and there was now Niels Bohr, who could even solve problems with 30 electrons or something like that." [From Heisenberg's lecture in Salam's book, pp. 92-93]

Bohr's Conjecture

"In the summer of 1922, Sommerfeld asked me whether I would be willing to follow him to a meeting at which Bohr would present his theory in Göttingen[...] There for the first time I learned how a man like Bohr worked on problems of atomic physics. When Bohr had given two of his lectures I dared once in a discussion to utter some criticism; I just mentioned some doubts, whether the formulae of Kramers which he had written on the blackboard could be exact

---

[159] Niels Bohr was a Danish physicist who contributed to the understanding of atomic structure and quantum theory. He won the Nobel Prize in Physics in 1922. The Bohr model of the atom proposed that energy levels of electrons are discrete and that the electrons revolve in stable orbits around the atomic nucleus but can jump from one energy level-orbit to another orbit. His underlying idea remains "valid" in today's physics. He lived from 1885-1962.

[...] he asked me for a long walk on the Hainberg near Göttingen to discuss the problem. I feel it was then that I felt I really learned what it means to work on an entirely new field in theoretical physics. **The first, for me quite shocking experience was that Bohr had calculated nothing. He had just guessed his results.** He knew the experimental situation in chemistry, he knew the valencies of the various atoms and he knew that his idea of the quantization of the orbits or rather his idea of the stability of the atom to be explained by the phenomenon of quantization, fitted somehow with the experimental situation in chemistry. **On this basis he simply guessed what he then gave us as his results.**" [From Heisenberg's lecture in Salam's book, pp. 93-94]

Author's comment: Scientists did not have super computers in Bohr's time to figure the high complexity of the electron orbits, so they **guessed** the answers. The orbits were based on a guess, not on observable evidence, not on a mathematical formula! The guess came before the formula. Even today super computers cannot figure out the problems of the movements of three bodies interrelating among themselves — the "three-body problem." There is no general solution to the problem, only restricted three-body problems that have *approximate* solutions. It is just too complex. How can scientists figure the complex and invisible movements of the electrons? The whole idea of an electron is built on a series of guesses.

" I asked him whether he really believed that one could derive these results by means of calculations based on classical mechanics. He said 'Well, I think that those classical pictures which I draw of the atoms are just as good as classical pictures can be' and he explained it in the following way. He said "we are now in a new field of physics, in which we know that the old concepts probably don't work. We see that they don't work, because otherwise atoms wouldn't be stable. On the other hand when we want to speak about atoms, we must use words and these words can only be taken from the old concepts, from the old language. Therefore we are in a hopeless dilemma, we are like sailors coming to a very far away country. They don't know the country and they see people whose language they have never heard, so they don't know how to communicate. Therefore, so far as the classical concepts work, that is, so far as we can speak about the motion of electrons, about their velocity, about their energy, etc., I think that my pictures are correct or at least I hope that they are correct, but nobody knows how far such a language goes'." [From Heisenberg's lecture in Salam's book, pp. 94-95]

Author's comment: The existence of atoms and electrons and their velocities are not much more than guesses because we cannot see them, we cannot measure them. Atoms and electrons are merely templates of what we think exists in the micro world.

No one has seen an atom, no one has seen an electron. Only by inference and theory can scientists "see" them. Even a scanning tunneling microscope cannot see atoms — they merely "feel" them by detecting their electronic pulses or currents and project a picture of the atom (electron cloud) onto a screen. Where does the picture of the atom come from? Software programs (algorithms) depict the picture of the atom as a cloud-like globe on a screen even though they are only measuring the electronic pulse or force. This is because the software tells the screen that atoms are globe like (and all the same size), since that is what scientists believe, not because the instrument actually sees them as globes of the same size. If the software was written differently the so-called atoms could be different sizes, different shapes, etc. Scientists (and their instruments) see the world through their theories, as Einstein said. See below for what Einstein said about this.

"When I came back from Copenhagen to Göttingen I decided that **I should again try to do some kind of guesswork there, namely to guess the intensities in the hydrogen spectrum. The Bohr theory didn't work well for these intensities. But why should it not be possible to guess them?**" [...] "**The formulae got too complicated** and there was no hope to get out anything. At the same time I also felt, if the mechanical system would be simpler, then it might be possible just to do the same things as Kramers and I had done in Copenhagen and **to guess the amplitudes.** ... It turned out that it really was quite simple to translate classical mechanics into quantum mechanics." [From Heisenberg's lecture in Salam's book, p. 96-97]

Author's comment: More guess work. This time pertaining to the intensities or amplitudes in the hydrogen spectrum. Again, we see that the formulas were too complex, so they guessed. And where did the quantum mechanics come from? It came from the classical mechanics, just like Einstein's theories were derived from the classical theories. Einstein merely used Lorentz's mathematical trick "from on high" so that the velocity of light would remain the same for all observers. Remember that Ptolemy's geocentric mathematical system worked even though its picture of the solar system was completely wrong. With the language of numbers you can do almost anything.

"Of course, classical theory would be included and quantum theory also would be included, but it was much too undefined and one had to add extra conditions [to the formula]. It turned out that one could replace the quantum conditions of Bohr's theory by a formula which was essentially equivalent to the sum rule of Thomas and Kuhn. **By adding such a condition one all of a sudden got into a consistent scheme. One could see that this set of assumptions**

worked, one could see that the energy was constant and so on. I was, however, not able to get a neat mathematical scheme out of it. **Very soon afterwards both Born and Jordan in Göttingen and Dirac**[160] **in Cambridge were able to invent a perfectly closed mathematical scheme; Dirac with very ingenious new methods on q numbers and Born and Jordan with more conventional methods on matrices."** [From Heisenberg's lecture in Salam's book, pp. 97-98]

Author's comment: As we see, with just a little adjustment, or by adding a special condition (*ad hoc*), they could fix a problem with their theory. Ptolemy did the same thing. He added special conditions to make his mistaken theory look correct. Why not look for a different theory? Why not question the theory so that you do not have to keep adding special conditions to make it work?

Einstein on Theory and Observation

"I told him [Einstein] that I did not believe any more in electronic orbits, in spite of the tracks in a cloud chamber. I felt that one should go back to those quantities which really can be observed and I also felt that this was just the kind of philosophy which he [Einstein] had used in relativity; because he also had abandoned absolute time and introduced only the time of the special coordinate system and so on. Well, he laughed at me and then he said 'but you just realize that it is completely wrong'. I answered: 'but why, is it not true that you have used this philosophy?' "Oh yes," he said, "I may have used it, but still it is nonsense!" Einstein explained to me that it was really the other way around. **He said "whether you can observe a thing or not depends on the theory which you use. It is the theory which decides what can be observed."** His argument was like this: "Observation means that we construct some connection between a phenomenon and our realization of the phenomenon. There is something happening in the atom, the light is emitted, the light hits the photographic plate, we see the photographic plate and so on and so on. In this whole course of events between the atom and your eye and your consciousness you must assume that everything works as in the old physics. If you change the theory concerning this sequence of events then of course the observation would be altered." **So he insisted that it was the theory which decides about what can be observed.** [From Heisenberg's lecture in Salam's book, pp. 99-100]

---

[160] Paul Adrien Maurice Dirac was an English theoretical physicist who was fundamental in the development of both quantum mechanics and quantum electrodynamics. Dirac shared the Nobel Prize in Physics for 1933 with Erwin Schrödinger for the discovery of new productive forms of atomic theory. He lived 1902-1984.

Author's comment: This is absolutely important to understand. Einstein understood this and we should also. Scientists indirectly observe atoms and electrons only because of the theory. You change the theory and you will observe things in a different way. Since theories are built upon theories, when you get one wrong, you in all probability get the others wrong. In today's astronomy, they have a very special view of the so-called red shift measured from the star's spectrum. If this is wrong, the great distances in space are wrong and the age of the universe is wrong. If Einstein's theories are wrong, then the theories on black holes and antimatter are wrong. So on and so forth. Real science is very limited; speculative science lives in a virtual world: "do not ask questions, just play the game."

"Mach himself had believed that the concept of the atom was only a point of convenience, a point of economy thinking, he didn't believe in the reality of the atoms. Nowadays everybody would say that this is nonsense, that it is quite clear that the atoms really exist." [From Heisenberg's lecture in Salam's book, p. 100]

Author's comment: The idea of atoms is an old idea and not necessarily reality. Real scientists must be open to new ideas.

"I should also add that when one has invented a new scheme which concerns certain observable quantities, then of course, the decisive question is: which of the old concepts can you really abandon? In the case of quantum theory it was more or less clear that you could abandon the idea of an electronic orbit." [From Heisenberg's lecture in Salam's book, p. 101]

Author's comment: Yet they still spoke and still speak of electronic orbits.

Rigorous and Dirty Mathematics

"I think that you understand now why I am always a bit skeptical of rigorous mathematical methods. Perhaps I should give a more serious reason for that: When you try too much for rigorous mathematical methods, you fix your attention on those points which are not important from the physics point and thereby you get away from the experimental situation. **If you try to solve a problem by rather dirty mathematics, as I have mostly done, then you are forced always to think of the experimental situation; and whatever formulae you write down, you try to**

compare the formulae with reality and thereby, somehow, you get closer to reality than by looking for the rigorous methods." [From Heisenberg's lecture in Salam's book, pp. 105-106]

Author's comment: The scientists try to find the perfect formulas to fit the observable evidence, that is, the "evidence" found through the perception of the theory. It is a good thing to try to fit the formulas with the evidence. But even here, they use "dirty mathematics" to attempt to do it and even so they do not really achieve it.

Abandoning Old Concepts

Also in Copenhagen we were not yet too happy about the interpretation because we felt that in the atom it seemed all right to abandon the concept of an electronic orbit. But what about in a cloud chamber? In a cloud chamber you see the electron moving along the track; is this an electronic orbit or not? [From Heisenberg's lecture in Salam's book, p. 110]

Author's comment: No electronic orbits? But what about the cloud chamber? What about it? It is also based on a huge amount of assumptions and they use the theory to "see" the electrons. Circular thinking!

Quantum Theory Understood

Bohr and I discussed these problems many, many nights and we were frequently in a state of despair. Bohr tried more in the direction of the duality between waves and particles; I preferred to start from the mathematical formalism and to look for a consistent interpretation. Finally Bohr went to Norway to think alone about the problem and I remained in Copenhagen. **Then I remembered Einstein's remark in our discussion. I remembered that Einstein had said that "It is the theory which decides what can be observed".** From there it was easy to turn around our question and not to ask "How can I represent in quantum mechanics this orbit of an electron in a cloud chamber?" but rather to ask "Is it not true that always only such situations occur in Nature, even in a cloud chamber, which can be described by the mathematical formalism of quantum mechanics?" By turning around I had to investigate what can be described in this formalism; and then it was very easily seen, especially when one used the new mathematical discoveries of Dirac and Jordan about transformation theory, that one could not describe at the same time the exact position and the exact velocity of an electron; **one had these uncertainty relations.** In this way things became clear. When Bohr returned to Copenhagen, he had found an

equivalent interpretation with his concept of complementarity, so finally we all agreed that now we had understood quantum theory. [From Heisenberg's lecture in Salam's book, p. 110-111]

Author's comment: The uncertainty principle in the quantum theory came about, not from evidence, but from the lack of it. They fabricated the idea of the so-called "uncertainty principle" to fit the lack of evidence.

Electrons and the Nucleus

Now I come to more recent developments. Perhaps I should, before I come to relativistic quantum theory, say a few words about nuclear physics. **The only point I want to make here again is that it is much easier to accept new concepts than to abandon old ones.** Actually, when the neutron was discovered by Chadwick in 1932, I think, it was almost trivial to say that the nucleus consists of protons and neutrons, but it was not quite so trivial to say that there are no electrons in the nucleus. The decisive point of those papers that I wrote about the structure of the nucleus was not that the nucleus consisted of protons and neutrons, but that in apparent contradiction to experiment there were not electrons in the nucleus. **Everybody up to that time had assumed that there must be electrons in the nucleus, because sometimes they come out, and it was rather odd to say that they have not been in the nucleus before. Of course, the idea was that the short-range forces between neutron and proton somehow might have to create electrons in the nucleus.** Anyway it seemed to be a good approximation to assume that such light particles cannot exist in the nucleus. I remember that I have been criticized very strongly for this assumption by extremely good physicists. I got one letter saying that it was really a scandal to assume that there were no electrons in the nucleus because one could just see them coming out; I would bring complete disorder into physics by such unreasonable assumptions and they could not understand my attitude. I just mentioned this small event, because it is really difficult to go away from something which seems so natural and so obvious that everybody had always accepted it. I think the greatest effort in the development of theoretical physics is always necessary at those points where one has to abandon old concepts. [From Heisenberg's lecture in Salam's book, pp. 112-114]

Author's comment: Logic be damned. Just because electrons seemed to come out of the nucleus of the atom, did not mean they were in it? Science must "abandon old concepts," if they do not fit the popular theory, even if the old concepts were logical?

Changing the Outlook of Atomic Physics

May I now turn to the problem of the elementary particles. I think that really the most decisive discovery in connection with the properties or the nature of elementary particles was the **discovery of antimatter by Dirac**. That was **an entirely new feature which apparently had to do with relativity,** with the replacement of the Galilei group by the Lorentz group. **I believe that this discovery of particles and antiparticles by Dirac has changed our whole outlook on atomic physics completely.** I do not know whether this change was realized at once at that time, probably it has been accepted only gradually; but I would like to explain why I consider it so fundamental.

We know from quantum theory that, for instance, a hydrogen molecule may consist of two hydrogen atoms or of one positive hydrogen ion and one negative hydrogen ion. Generally one can say that every state consists virtually of all possible configurations by which you can realize the same kind of symmetry. Now as soon as one knows that one can create pairs according to Dirac's theory, then one has to consider an elementary particle as a compound system; because virtually it could be this particle plus a pair or this particle plus two pairs and so on, and **so all of a sudden the whole idea of an elementary particle has changed.** Up to that time I think every physicist had thought of the elementary particles along the line of the philosophy of Democritus, namely by considering these elementary particles are unchangeable units which are just given in nature and are just always the same thing, they never change, they never can be transmuted into anything else. They are not dynamical systems, they just exist in themselves.

After Dirac's discovery everything looked different, because now one could ask, why should a proton be only a proton, why should a proton not sometimes be a proton plus a pair of one electron and one positron and so on. [From Heisenberg's lecture in Salam's book, pp. 114-115]

Author's comment: We see that the idea of antimatter came from relativity. They needed some duality of matter to make their theory work, so they just invented antimatter. Another *ad hoc* addition to a theory that was not working. Maybe they should just toss the theory and look for something better, or maybe they should just look at the idea of "light's constant velocity" in a different light as we have done in this book.

Pair Creation

Either you can divide matter again and again into smaller and smaller bits or you cannot divide matter up to infinity and then you come to smallest particles. Now all of a sudden we saw a third possibility: we can divide matter again and again but we never get to smaller particles because we must create particles by energy, by kinetic energy, and since we have pair creation this can go on forever [...]

One had great difficulties in avoiding infinities in quantum electrodynamics and, more generally, in quantum field theory and with interaction. **I agree completely with Dirac in disliking infinities in the sense that if you introduce infinity in physics, you just talk nonsense, it cannot be done. Therefore I tried to think of mathematical schemes in which you can avoid infinities** from the very beginning. [...] and so it was natural to come to the SW-matrix and to say that the S-matrix or scattering matrix is a rational basis for a theory.

Again of course it is considerably easier to go this first step, namely to say that such and such things can be observed, then to go the next step and to narrow down the assumptions. But finally you have to make new assumptions and end up with saying "Such and such things cannot be observed anymore." So the question was now "How can we narrow down the concept of the S-matrix in order to get something which is really workable, in which we can define what we mean, in which we can formulate natural laws."[...]

Therefore I tried to narrow the scheme down by saying "There shall be local field operators but these operators may work in a Hilbert space which does not have an ordinary metric but may have an indefinite metric." **The advantage of this scheme was that one could actually avoid infinities but, of course, at the very high price, namely at the price of loosing the definite metric in Hilbert space.** On the other hand, by that time the whole scheme already looked rather convincing to me, because the experiments in the meantime had proved that there was actually multiple production of particles. [From Heisenberg's lecture in Salam's book, pp. 116-119]

Author's comment: Because the scientists did not like the idea of dividing matter into smaller and smaller parts to infinity, they instead invented the "creation" of matter by using a "scheme" that avoided infinities (½ of ½ of ½). Instead of inventing such an artificial scheme, maybe they should have looked at their misperception and misuse of mathematics.

Pauli's Critical Acumen
"Bohr had dared [to] publish papers which he could not prove and which were right after all." [From Heisenberg's lecture in Salam's book, p. 121]

Author's comment: Whether it was proved or not, I do not know. Yes, sometimes a guess might work. This happens, because theories are just guesses. What I am trying to point out is that science is not that exact science it pretends to be. Real scientists know this. The professed scientists do not understand this. And many in academia know a lot of what they do is just pretend — "play the game" and keep your job.

The Following Taken From Paul Dirac's 1968 Lecture:

Cosmological Speculation

One field of work in which there has been too much speculation is cosmology. There are very few hard facts to go on, but theoretical workers have been busy constructing various models for the universe, based on any assumptions that they fancy. These models are probably all wrong. It is usually assumed that the laws of nature have always been the same as they are now. There is no justification for this. The laws may be changing, and in particular quantities which are considered to be constants of nature may be varying with cosmological time. Such variations would completely upset the model makers. [From Dirac's lecture in Salam's book, pp. 127-128]

Author's comment: I agree with this statement. There is no real proof that the laws of nature have always been the same and will always be the same. Yet this seems reasonable to many, which may have something to do with mankind's belief in a supreme being who make made the universe in harmony and order. It should seem more reasonable to the evolutionists if the laws varied. Yet most scientists think that the laws were always the same. Strange.

Theoretical physicists accept the need for mathematical beauty as an act of faith. There is no compelling reason for it, but it has proved a very profitable objective in the past. For example, the main reason why the theory of relativity is so universally accepted is its mathematical beauty. [From Dirac's lecture in Salam's book, p. 128]

Author's comment: Ptolemy's geocentric theory was overly complex, while the heliocentric model was more simple and beautiful. The classical theory of relativity also was mathematically beautiful *and* without paradoxes, multiple "local" times and different lengths in different reference systems.

Impact of Relativity

In order to understand the atmosphere in which theoretical physicists were then working, one must appreciate the enormous influence of relativity. [...] The excitement was quite unprecedented in the history of science.

Against this background of excitement, physicists were trying to understand the mystery of the stability of atoms. Schrödinger, like everyone else, was caught up by the new ideas, and so he tried to set up a quantum mechanics with the framework of relativity. Everything had to be expressed in terms of vectors and tensors in space-time. [...]

Schrödinger was working from a beautiful idea of de Broglie connecting waves and particles in a relativistic way. De Broglie's idea applied only to free particles, and Schrödinger tried to generalize it to an electron bound in an atom. Eventually he succeeded, keeping within the relativistic framework. **But when he applied his theory to the hydrogen atom, he found it did not agree with experiment. The discrepancy was due to his not having taken the spin of the electron into account.** It was not then known. [From Dirac's lecture in Salam's book, pp. 131-132]

Author's comment: Again we see how the theory of relativity had entered into modern science and stirred matters. They tried to keep their theories within the relativistic framework. To do that they needed the *ad hoc* theory of "the spin of the electron." (The "spin" of the electron is not really a spin, just a misnomer used to describe a mathematical aspect of the theory.) I ask, why not question the theory, and not add special conditions to it to make it work better with relativity. Why not get rid of Einstein's relativity and see the world more as it is.

On the Wrong Track?

"It involved setting up rules for discarding the infinities, rules which are precise, so as to leave well-defined residues that can be compared with experiment. But still one is using working rules and not regular mathematics." [From Dirac's lecture in Salam's book, p. 138]

Author's comment: Special rules or "working rules" are just *ad hoc* rules.

"As we cannot now use the Schrödinger picture, we cannot use the regular physical interpretation of quantum mechanics... **we have to feel our way** towards a new physical interpretation which can be used with the Heisenberg picture." [From Dirac's lecture in Salam's book, p. 141]

Author's comment: "Feel" our way?

One finds that the parameters m and e denoting the mass and charge of the electron in the starting equations are not the same as the observed values for these quantities. If we keep the symbols m and e to denote the observed values, we have to replace the m and and e in the starting equations by $m + \delta m$ and $e + \delta e$, where δm and δe are small corrections which can be calculated. **This procedure is known as renormalization."**[161] [From Dirac's lecture in Salam's book, p. 141]

Author's comment: More small corrections mean more *ad hoc* corrections, which they call "renormalization."

Difficulty in Quantum Electrodynamics

Such a change in the starting equations is permitted. **We can take any starting equations we like, and then develop the theory by making deductions from them.** You might think that the work of the theoretical physicist is easy if he can make any starting assumptions he likes, but the difficulty arises because he needs the same starting assumptions for all the applications of the theory. This very strongly restricts his freedom. **Renormalization is permitted because it is a simple change which can be applied universally whenever one has charged particles interacting with the electromagnetic field. [...]**

The ultimate goal is to obtain suitable starting equations from which the whole of atomic physics can be deduced. We are still far from it. [From Dirac's lecture in Salam's book, p. 142]

Author's comment: Yes we are far from that. In reality a beautiful, simple equation of *everything* may be just fantasy.

---

[161] "Renormalization" is just another *ad hoc* addition to a theory with problems.

— End of critique and comments on the lectures in Abdus Salam's book —

Basic Assumptions of Quantum Theory are Arbitrary

Norwood Russel Hanson, a Cambridge professor of philosophy, breaks down some of the assumptions in the elementary particle theory in his book entitled, *Patterns of Discovery: An Inquiry into the Conceptual Foundation of Science*. He says in his Introduction,

> "*Observation, evidence, facts*; these notions, if drawn from the 'catalogue-sciences' of school and undergraduate text-books, will ill prepare one for understanding foundations of particle theory."[162]

All Electrons Are Identical?

No, your undergraduate studies in science will not give you the fundamental assumptions of science, which are in many cases only educated guesses. Hanson's book is full of the author's observations about the arbitrary assumptions of the quantum theory. I list one such arbitrary assumption here — that all electrons are identical:

> It is an indispensable condition of quantum theory that all electrons, all protons, all neutrons, must be identical; the successes of microphysics rest on this conception. If it is questioned, then all the achievements of two generations come under question as well. [Ibid., Hanson, pp. 133-134]

> The power and success of quantum theory consists in the pattern of interlocked, systematic accounts it gives of the behaviour of complex bodies. Since it does this only by postulating the absolute identity [sameness] of all elementary particles of the same type, what better reason could there be for saying that all elementary particles of the same type are identical? 'Then the identity of all electrons is just an assumption. It is true because physics will not hear of it being false. **It is a definition pure and simple, hence arbitrary**.' [...] Indeed, without this conception experiments would not even make sense. All the data, the facts, the observations, bear the stamp of this unifying conception. [Ibid., Hanson, p. 134, Emphasis added] [...]

> Why are all the electrons identical? Because the world as we now know it becomes intelligible by supposing these things to be the case. What better reason for saying that they are the case? [Ibid., Hanson, p. 134]

---

[162] N. R. Hanson, *Patterns of Discovery: An Inquiry into the Conceptual Foundation of Science,* Cambridge University Press, 1958, pp. 1-2. Emphasis in text.

This concept of electrical breakdown underlies all explanations of why electrons are, and must be, the same size. [163] [Ibid., Hanson, p. 135]

Scientists Observe Through Their Theories

Einstein put it this way:

"Oh yes," he [Einstein] said, "I may have used it, but still it is nonsense."

Einstein explained to me[164] that it was really the other way around. He said **"whether you can observe a thing or not depends on the theory which you use. It is the theory which decides what can be observed."** [...]

"So he [Einstein] insisted that it was the theory which decides about what can be observed."[165]

Hanson said it this way:

Physical theories provide patterns within which data appear intelligible. [...] A theory is not pieced together from observed phenomena; it is rather what makes it possible to observe phenomena...."[166]

In the theory of elementary particles, for sure, it is the theory you use that determines what you see, because you cannot see or observe these particles. Yet, the theory itself is nothing but a set of arbitrary assumptions.

Constancy in Science Was Often Only Assumptions
Light's Velocity in a Vacuum Always the Same?

Light's velocity, in all the spectrum, is assumed to be about 300,000 kilometers per second in a vacuum (or 'exactly' 299,792 km/s). The *velocity* of light is believed to be a fundamental constant and law of the universe, except it has <u>not</u> been measured outside our solar system,[167] and has only been measured by two way (back and forth) methods.[168] Also, because of

---

[163] N.R. Hanson, *Patterns of Discovery: An Inquiry into the Conceptual Foundation of Science,* Cambridge University Press, 1958. Quotes taken from pp. 133-135.

[164] "me" here refers to Werner Heisenberg.

[165] Heisenberg's 1968 Lecture included in: Abdus Salam, *Unification of Fundamental Forces: The First of the 1988 Dirac Memorial Lectures*, Cambridge University Press, Cambridge, GB, 1990, p. 99-100.

[166] Ibid., Hanson, p. 90.

[167] Depending on the nature of light (energy-source) each star system may have a different value to the velocity of light, if the velocity of light has any relationship to the energy of each star's emission process. See xxxx

[168] https://en.wikipedia.org/wiki/One-way_speed_of_light As of today their has been no on-way speed of light measurement because the problem of synchronizing the clocks at the source and receiver positions.

Einstein's equations, the velocity of light is believed to be the speed limit in the universe — nothing can go faster than light. The theory of relativity is the genesis of the idea that the velocity of light is a constant in nature. Yet, as we will see, it is not now a constant in our solar system or the universe, nor is there any proof that it was in the past or that it will be in the future. Paul Dirac, considered one of the greatest physicists of the twentieth century, said the following in a lecture:

> One field of work in which there has been too much speculation is cosmology. There are very few hard facts to go on, but theoretical workers have been busy constructing various models for the universe, based on any assumptions that they fancy. These models are probably all wrong. **It is usually assumed that the laws of nature have always been the same as they are now. There is no justification for this. The laws may be changing,** and in particular quantities which are considered to be constants of nature may be varying with cosmological time.[169]

Constant Decay Rates?

Furthermore it was believed at first that these decay rates were constant.

> "Radioactivity was discovered by Becquerel in 1896. In 1906, Millikan stated, 'Radioactivity has been found to be independent of all physical as well as chemical conditions. The lowest cold or greatest heat does not appear to affect it in the least. Radioactivity seems to be as unalterable a property of the atoms of radioactive substances, as is weight itself.' This state of mind established the modern view, which is quite generally held today.... The electroscope and spinthariscope were used in early study of radioactive alpha decay rates. The inherent limitations of these early instruments led to erroneous conclusions:
>
> 1. That radioactive decay rates are constant.
>
> 2. That these rates cannot be altered by change of the energy state of the electrons orbiting the nucleus.
>
> 3. That radioactivity results from processes which involve only the atomic nucleus.
>
> Refinements in electronics resulted in the development of sophisticated counting apparatus. The equipment was used in the demonstration by several investigators (1949-73) of rather easily induced changes in the disintegration rates of 14 radionuclides, including $_{14}C$, $_{60}Co$, and $_{137}Cs$. The observed variations in the decay rates, (changes in the half life) were produced by changes in pressure, temperature, chemical state, electric potential stress of monomolecular layers, etc. ... The decay 'constant' is now considered to be a variable." [H.C. Dudley, "Is There Ether?,"Industrial Research, Nov

---

[169] Paul Adrian Maurice Dirac, "Methods in Theoretical Physics," *Unification of Fundamental Forces*, by Abdus Salam, Cambridge University Press 1990. Emphasis added.

15, 1974, p. 42; my emphasis]ial Research, Nov 15, 1974, p. 42; my emphasis] [170]

Constant Atomic Clock Rates?

Atomic clocks even seem to change their rates of decay by the direction in which they travel in an airplane. Those going westward gained time; those going eastward lost time (Hafele, Keating, 1972, "Around-the world atomic clocks," **Science** 177 [4044]).

My Conclusion about the Guessing Game

Much has been gained when science is honest with itself, and not blinded by biased mindsets and dogma. The mechanical, electrical, electronic and nana-technology fields, to name a few, can be thankful for the science of physical observation. On the other hand, astronomy, the *new* physics and evolutional theories have become engrossed in theory and mathematics not backed by real observation for over 175 years. Astronomy and evolution do not have any witnesses that observed what happened in the billions of years of their theories, except through their arbitrary guesses and mathematics. I love real science based on sound observation, not science based on mere conjecture and historical guesses.

The question: is modern science any better than philosophy?

The answer will depend on which science you are referring to. Are we referring to historical science that studies past ages where we have no witnesses? Are we talking about the science of things we cannot see, such as black holes or elementary particles? Are we speaking about the science pertaining to the future (weather, earthquakes, etc.) where our view is limited by our arbitrary and guess-work theories? Or are we referring to the science of what we can observe with our senses, or with reliable indirect methods, and does this science use mathematics in a trustworthy manner?

---

[170] H.C. Dudley, "Is There Ether?," *Industrial Research*, Nov 15, 1974, p. 42

Appendix 8: Why is There Anything

— Footnote reference reprint —

[Grok search results, Aug, 2025]
https://x.com/i/grok?conversation=1944215194504221069

My Question:
Why is there anything in the Universe. Why? Can something come from nothing.

Grok answers:

In these questions are we dealing with definition of "nothing" or what?

The question "Why is there anything in the universe?"—often phrased as "Why is there something rather than nothing?"—is one of the most profound in philosophy and science, dating back to thinkers like Leibniz in the 17th century. It's compounded by related queries: Can something emerge from nothing? And yes, much of the debate does revolve around the definition of "nothing," as we'll explore. I'll break this down step by step, drawing on philosophical arguments, modern physics, and cosmology, while addressing potential misconceptions.

The Role of "Nothing" in These Questions

At the heart of your query is indeed the concept of "nothing." In everyday language, "nothing" means absolute absence—no matter, energy, space, time, laws of physics, or even potential for existence. But this pure philosophical nothing is tricky: if "nothing" truly has no properties, it can't "exist" in any meaningful way, leading some to argue that something must exist by necessity, as nothingness would be unstable or impossible. For instance, defining nothing as having "no size, no shape, no properties" paradoxically gives it properties (like the property of having no properties), making true nothing inconceivable or self-contradictory.

In physics, however, "nothing" is redefined pragmatically as the vacuum —the lowest energy state of a system, like empty space devoid of particles. But this isn't absolute emptiness: quantum mechanics shows the vacuum teems with virtual particles popping in and out due to quantum fluctuations, governed by the Heisenberg uncertainty principle. It also has vacuum energy, a positive value linked to the cosmological constant, which drives the universe's accelerated expansion. So, when physicists discuss "something from nothing," they usually mean from this quantum vacuum, not philosophical nothingness. This distinction is key: debates often blur when people equate the two, leading to confusion about whether science "proves" creation from true nothing (it doesn't, as the vacuum presupposes existing quantum fields and laws).

Philosophical Perspectives on Why There's Something

Philosophically, answers vary, but none are definitive proofs— they're more like reasoned possibilities:

- **Necessity or Inevitability**: Some argue existence is a brute fact; there must be something because nothingness isn't a stable or possible state. For example, if nothing existed, there'd be no reason for change, so something is inevitable. Others invoke modal logic: in all possible worlds, something exists because "nothing" isn't a world at all.
- **Theological Views**: Many traditions posit a creator (e.g., God) as the reason, arguing that contingent existence (things that could not exist) requires a necessary being to explain it. This avoids infinite regress but introduces faith-based elements.
- **Absurdity or Unanswerability**: Existentialists like Heidegger saw it as the fundamental question of metaphysics, but some modern philosophers (e.g., influenced by Wittgenstein) say it's meaningless, as "why" presupposes causation within existence. Or, as physicist Sean Carroll puts it, the universe might just be a self-contained system with no external "why."

These views don't rely on empirical evidence but on logic and ontology (the study of being).

Scientific Perspectives: Can Something Come from Nothing?

Science doesn't claim the universe came from absolute nothing but offers models for its origin from a prior state. The Big Bang theory, the leading framework, describes the universe expanding from a hot, dense singularity about 13.8 billion years ago—not "from nothing," but from an initial condition where space, time, and matter emerged. The singularity isn't "nothing"; it's a breakdown in our equations (general relativity fails there), possibly resolved by quantum gravity theories.

On whether something can emerge from "nothing" (quantum vacuum):

- **Quantum Fluctuations**: Yes, in a limited sense. The Schwinger effect and Casimir effect show particles can arise from vacuum under strong fields, conserving energy via borrowing from uncertainty. Virtual particles constantly flicker in and out, and in cosmology, quantum fluctuations during inflation (rapid early expansion) seeded galaxy formation. But this requires pre-existing space-time and laws— not true nothing.

- **Universe from Vacuum**: Models like those from Lawrence Krauss suggest the universe arose as a quantum fluctuation in a multiverse's vacuum, where total energy is zero (gravity balances positive energy). This "nothing" has structure, though.

Critics note this doesn't violate conservation laws but doesn't explain the laws themselves. The first law of thermodynamics says energy can't be created from nothing, but quantum mechanics allows temporary violations.

Recent Theories (as of 2025) on the Universe's Origin

Cosmology evolves, with alternatives to a strict "from nothing" Big Bang:

Theory	Key Idea	Implications for "Something from
Bouncing Cosmology	Universe cycles through contractions and expansions ("bounces"), avoiding a singularity. Recent models	No absolute beginning; "something" always existed in cycles, sidestepping true
Black Hole Universe	Our universe formed inside a black hole from a prior collapsing one, with the	Origin from a prior "something" (collapsing matter), not nothing.
Eternal Inflation/ Multiverse	Inflation never fully stops; our universe is one bubble in an eternally inflating multiverse.	No global beginning; "something" (the multiverse) is eternal.
Quantum Gravity Approaches (e.g.,	Replace singularity with a quantum "bounce" or higher dimensions.	Universe from a pre-Big Bang quantum state, not nothing. Still speculative, but

As of mid-2025, no theory claims creation from absolute nothing; they posit prior states or eternal processes. Evidence like cosmic microwave
background supports expansion from a hot dense state, but the ultimate "why" remains open—perhaps unanswerable scientifically, as it probes beyond observable reality.

In summary, the "why" might be necessity, chance, or design, but science shows the universe likely didn't spring from pure nothing. Instead, "nothing" in physics is something, and origins trace to quantum or cyclic processes. If this feels unsatisfying, it's because these questions push the limits of human inquiry.

Flashlight's/Searchlight's Diminishing Power: Energy = Distance

No, you could not see the light from a typical handheld flashlight shone from the Moon to Earth with the naked eye, as the beam would diverge too much over the ~384,000 km distance, resulting in an illuminance on Earth far below the human eye's detection threshold (roughly equivalent to a visual magnitude of 12–14 or dimmer, while the naked-eye limit is around magnitude 6–7 in dark skies).

However, it could be visible under specific conditions with a highly powerful and well-collimated light source, such as a WWII-era carbon arc searchlight (e.g., those with ~800 million candela intensity):

- **Powerful, collimated source**: The light must have extremely high intensity (hundreds of millions of candela) and low beam divergence (around 1 degree full angle or less, achievable with large-aperture searchlights) to deliver detectable illuminance (~10^{-9} lux or higher) at Earth, appearing as a point source with a visual magnitude of ~6.5–7.
- **Dark phase of the Moon**: The flashlight must be on the Earth-facing side during a new moon (when the near side is unilluminated by the Sun) to avoid the Moon's surface brightness overwhelming the signal.
- **Optimal viewing on Earth**: Nighttime, clear atmosphere, no light pollution, dark-adapted eyes, and ideally in a very dark location (e.g., remote area with Bortle scale 1–2 skies). Binoculars or a telescope would make detection easier, but naked-eye visibility is possible at the upper end of these parameters.
- **No obstructions**: The beam must be precisely aimed at the observer's location on Earth, with no interference from Earth's atmosphere scattering the light excessively (though minimal at these intensities).

Even then, the light would appear as a faint, star-like point in the Moon's position in the sky, not as a visible beam or illuminated spot on the ground, due to the vacuum of space preventing side-scattering. For reference, the beam spreads significantly but remains directional if well-collimated

ON THE ELECTRODYNAMICS OF MOVING BODIES

By A. EINSTEIN

June 30, 1905

It is known that Maxwell's electrodynamics—as usually understood at the present time—when applied to moving bodies, leads to asymmetries which do not appear to be inherent in the phenomena. Take, for example, the reciprocal electrodynamic action of a magnet and a conductor. The observable phenomenon here depends only on the relative motion of the conductor and the magnet, whereas the customary view draws a sharp distinction between the two cases in which either the one or the other of these bodies is in motion. For if the magnet is in motion and the conductor at rest, there arises in the neighbourhood of the magnet an electric field with a certain definite energy, producing a current at the places where parts of the conductor are situated. But if the magnet is stationary and the conductor in motion, no electric field arises in the neighbourhood of the magnet. In the conductor, however, we find an electromotive force, to which in itself there is no corresponding energy, but which gives rise—assuming equality of relative motion in the two cases discussed—to electric currents of the same path and intensity as those produced by the electric forces in the former case.

Examples of this sort, together with the unsuccessful attempts to discover any motion of the earth relatively to the "light medium," suggest that the phenomena of electrodynamics as well as of mechanics possess no properties corresponding to the idea of absolute rest. They suggest rather that, as has already been shown to the first order of small quantities, the same laws of electrodynamics and optics will be valid for all frames of reference for which the equations of mechanics hold good.[1] We will raise this conjecture (the purport of which will hereafter be called the "Principle of Relativity") to the status of a postulate, and also introduce another postulate, which is only apparently irreconcilable with the former, namely, that light is always propagated in empty space with a definite velocity c which is independent of the state of motion of the emitting body. These two postulates suffice for the attainment of a simple and consistent theory of the electrodynamics of moving bodies based on Maxwell's theory for stationary bodies. The introduction of a "luminiferous ether" will prove to be superfluous inasmuch as the view here to be developed will not require an "absolutely stationary space" provided with special properties, nor

---

[1] The preceding memoir by Lorentz was not at this time known to the author.

assign a velocity-vector to a point of the empty space in which electromagnetic processes take place.

The theory to be developed is based—like all electrodynamics—on the kinematics of the rigid body, since the assertions of any such theory have to do with the relationships between rigid bodies (systems of co-ordinates), clocks, and electromagnetic processes. Insufficient consideration of this circumstance lies at the root of the difficulties which the electrodynamics of moving bodies at present encounters.

I. KINEMATICAL PART

§ 1. Definition of Simultaneity

Let us take a system of co-ordinates in which the equations of Newtonian mechanics hold good.[2] In order to render our presentation more precise and to distinguish this system of co-ordinates verbally from others which will be introduced hereafter, we call it the "stationary system."

If a material point is at rest relatively to this system of co-ordinates, its position can be defined relatively thereto by the employment of rigid standards of measurement and the methods of Euclidean geometry, and can be expressed in Cartesian co-ordinates.

If we wish to describe the *motion* of a material point, we give the values of its co-ordinates as functions of the time. Now we must bear carefully in mind that a mathematical description of this kind has no physical meaning unless we are quite clear as to what we understand by "time." We have to take into account that all our judgments in which time plays a part are always judgments of *simultaneous events*. If, for instance, I say, "That train arrives here at 7 o'clock," I mean something like this: "The pointing of the small hand of my watch to 7 and the arrival of the train are simultaneous events."[3]

It might appear possible to overcome all the difficulties attending the definition of "time" by substituting "the position of the small hand of my watch" for "time." And in fact such a definition is satisfactory when we are concerned with defining a time exclusively for the place where the watch is located; but it is no longer satisfactory when we have to connect in time series of events occurring at different places, or—what comes to the same thing—to evaluate the times of events occurring at places remote from the watch.

We might, of course, content ourselves with time values determined by an observer stationed together with the watch at the origin of the co-ordinates, and co-ordinating the corresponding positions of the hands with light signals, given out by every event to be timed, and reaching him through empty space. But this co-ordination has the disadvantage that it is not independent of the standpoint of the observer with the watch or clock, as we know from experience.

---

[2]i.e. to the first approximation.

[3]We shall not here discuss the inexactitude which lurks in the concept of simultaneity of two events at approximately the same place, which can only be removed by an abstraction.

We arrive at a much more practical determination along the following line of thought.

If at the point A of space there is a clock, an observer at A can determine the time values of events in the immediate proximity of A by finding the positions of the hands which are simultaneous with these events. If there is at the point B of space another clock in all respects resembling the one at A, it is possible for an observer at B to determine the time values of events in the immediate neighbourhood of B. But it is not possible without further assumption to compare, in respect of time, an event at A with an event at B. We have so far defined only an "A time" and a "B time." We have not defined a common "time" for A and B, for the latter cannot be defined at all unless we establish *by definition* that the "time" required by light to travel from A to B equals the "time" it requires to travel from B to A. Let a ray of light start at the "A time" t_A from A towards B, let it at the "B time" t_B be reflected at B in the direction of A, and arrive again at A at the "A time" t'_A.

In accordance with definition the two clocks synchronize if

$$t_B - t_A = t'_A - t_B.$$

We assume that this definition of synchronism is free from contradictions, and possible for any number of points; and that the following relations are universally valid:—

1. If the clock at B synchronizes with the clock at A, the clock at A synchronizes with the clock at B.

2. If the clock at A synchronizes with the clock at B and also with the clock at C, the clocks at B and C also synchronize with each other.

Thus with the help of certain imaginary physical experiments we have settled what is to be understood by synchronous stationary clocks located at different places, and have evidently obtained a definition of "simultaneous," or "synchronous," and of "time." The "time" of an event is that which is given simultaneously with the event by a stationary clock located at the place of the event, this clock being synchronous, and indeed synchronous for all time determinations, with a specified stationary clock.

In agreement with experience we further assume the quantity

$$\frac{2AB}{t'_A - t_A} = c,$$

to be a universal constant—the velocity of light in empty space.

It is essential to have time defined by means of stationary clocks in the stationary system, and the time now defined being appropriate to the stationary system we call it "the time of the stationary system."

§ 2. On the Relativity of Lengths and Times

The following reflexions are based on the principle of relativity and on the principle of the constancy of the velocity of light. These two principles we define as follows:—

1. The laws by which the states of physical systems undergo change are not affected, whether these changes of state be referred to the one or the other of two systems of co-ordinates in uniform translatory motion.

2. Any ray of light moves in the "stationary" system of co-ordinates with the determined velocity c, whether the ray be emitted by a stationary or by a moving body. Hence

$$\text{velocity} = \frac{\text{light path}}{\text{time interval}}$$

where time interval is to be taken in the sense of the definition in § 1.

Let there be given a stationary rigid rod; and let its length be l as measured by a measuring-rod which is also stationary. We now imagine the axis of the rod lying along the axis of x of the stationary system of co-ordinates, and that a uniform motion of parallel translation with velocity v along the axis of x in the direction of increasing x is then imparted to the rod. We now inquire as to the length of the moving rod, and imagine its length to be ascertained by the following two operations:—

(a) The observer moves together with the given measuring-rod and the rod to be measured, and measures the length of the rod directly by superposing the measuring-rod, in just the same way as if all three were at rest.

(b) By means of stationary clocks set up in the stationary system and synchronizing in accordance with § 1, the observer ascertains at what points of the stationary system the two ends of the rod to be measured are located at a definite time. The distance between these two points, measured by the measuring-rod already employed, which in this case is at rest, is also a length which may be designated "the length of the rod."

In accordance with the principle of relativity the length to be discovered by the operation (a)—we will call it "the length of the rod in the moving system"— must be equal to the length l of the stationary rod.

The length to be discovered by the operation (b) we will call "the length of the (moving) rod in the stationary system." This we shall determine on the basis of our two principles, and we shall find that it differs from l.

Current kinematics tacitly assumes that the lengths determined by these two operations are precisely equal, or in other words, that a moving rigid body at the epoch t may in geometrical respects be perfectly represented by *the same* body *at rest* in a definite position.

We imagine further that at the two ends A and B of the rod, clocks are placed which synchronize with the clocks of the stationary system, that is to say that their indications correspond at any instant to the "time of the stationary system" at the places where they happen to be. These clocks are therefore "synchronous in the stationary system."

We imagine further that with each clock there is a moving observer, and that these observers apply to both clocks the criterion established in § 1 for the synchronization of two clocks. Let a ray of light depart from A at the time[4] t_A,

---

[4] "Time" here denotes "time of the stationary system" and also "position of hands of the moving clock situated at the place under discussion."

let it be reflected at B at the time t_B, and reach A again at the time t'_A. Taking into consideration the principle of the constancy of the velocity of light we find that

$$t_B - t_A = \frac{r_{AB}}{c - v} \text{ and } t'_A - t_B = \frac{r_{AB}}{c + v}$$

where r_{AB} denotes the length of the moving rod—measured in the stationary system. Observers moving with the moving rod would thus find that the two clocks were not synchronous, while observers in the stationary system would declare the clocks to be synchronous.

So we see that we cannot attach any *absolute* signification to the concept of simultaneity, but that two events which, viewed from a system of co-ordinates, are simultaneous, can no longer be looked upon as simultaneous events when envisaged from a system which is in motion relatively to that system.

§ 3. Theory of the Transformation of Co-ordinates and Times from a Stationary System to another System in Uniform Motion of Translation Relatively to the Former

Let us in "stationary" space take two systems of co-ordinates, i.e. two systems, each of three rigid material lines, perpendicular to one another, and issuing from a point. Let the axes of X of the two systems coincide, and their axes of Y and Z respectively be parallel. Let each system be provided with a rigid measuring-rod and a number of clocks, and let the two measuring-rods, and likewise all the clocks of the two systems, be in all respects alike.

Now to the origin of one of the two systems (k) let a constant velocity v be imparted in the direction of the increasing x of the other stationary system (K), and let this velocity be communicated to the axes of the co-ordinates, the relevant measuring-rod, and the clocks. To any time of the stationary system K there then will correspond a definite position of the axes of the moving system, and from reasons of symmetry we are entitled to assume that the motion of k may be such that the axes of the moving system are at the time t (this "t" always denotes a time of the stationary system) parallel to the axes of the stationary system.

We now imagine space to be measured from the stationary system K by means of the stationary measuring-rod, and also from the moving system k by means of the measuring-rod moving with it; and that we thus obtain the co-ordinates x, y, z, and ξ, η, ζ respectively. Further, let the time t of the stationary system be determined for all points thereof at which there are clocks by means of light signals in the manner indicated in § 1; similarly let the time τ of the moving system be determined for all points of the moving system at which there are clocks at rest relatively to that system by applying the method, given in § 1, of light signals between the points at which the latter clocks are located.

To any system of values x, y, z, t, which completely defines the place and time of an event in the stationary system, there belongs a system of values ξ,

η, ζ, τ, determining that event relatively to the system k, and our task is now to find the system of equations connecting these quantities.

In the first place it is clear that the equations must be *linear* on account of the properties of homogeneity which we attribute to space and time.

If we place $x' = x - vt$, it is clear that a point at rest in the system k must have a system of values x', y, z, independent of time. We first define τ as a function of x', y, z, and t. To do this we have to express in equations that τ is nothing else than the summary of the data of clocks at rest in system k, which have been synchronized according to the rule given in § 1.

From the origin of system k let a ray be emitted at the time τ_0 along the X-axis to x', and at the time τ_1 be reflected thence to the origin of the co-ordinates, arriving there at the time τ_2; we then must have $\frac{1}{2}(\tau_0 + \tau_2) = \tau_1$, or, by inserting the arguments of the function τ and applying the principle of the constancy of the velocity of light in the stationary system:—

$$\frac{1}{2}\left[\tau(0,0,0,t) + \tau\left(0,0,0,t + \frac{x'}{c-v} + \frac{x'}{c+v}\right)\right] = \tau\left(x',0,0,t + \frac{x'}{c-v}\right).$$

Hence, if x' be chosen infinitesimally small,

$$\frac{1}{2}\left(\frac{1}{c-v} + \frac{1}{c+v}\right)\frac{\partial\tau}{\partial t} = \frac{\partial\tau}{\partial x'} + \frac{1}{c-v}\frac{\partial\tau}{\partial t},$$

or

$$\frac{\partial\tau}{\partial x'} + \frac{v}{c^2 - v^2}\frac{\partial\tau}{\partial t} = 0.$$

It is to be noted that instead of the origin of the co-ordinates we might have chosen any other point for the point of origin of the ray, and the equation just obtained is therefore valid for all values of x', y, z.

An analogous consideration—applied to the axes of Y and Z—it being borne in mind that light is always propagated along these axes, when viewed from the stationary system, with the velocity $\sqrt{c^2 - v^2}$ gives us

$$\frac{\partial\tau}{\partial y} = 0, \frac{\partial\tau}{\partial z} = 0.$$

Since τ is a *linear* function, it follows from these equations that

$$\tau = a\left(t - \frac{v}{c^2 - v^2}x'\right)$$

where a is a function $\phi(v)$ at present unknown, and where for brevity it is assumed that at the origin of k, $\tau = 0$, when $t = 0$.

With the help of this result we easily determine the quantities ξ, η, ζ by expressing in equations that light (as required by the principle of the constancy of the velocity of light, in combination with the principle of relativity) is also propagated with velocity c when measured in the moving system. For a ray of light emitted at the time $\tau = 0$ in the direction of the increasing ξ

$$\xi = c\tau \text{ or } \xi = ac\left(t - \frac{v}{c^2 - v^2}x'\right).$$

But the ray moves relatively to the initial point of k, when measured in the stationary system, with the velocity $c - v$, so that

$$\frac{x'}{c - v} = t.$$

If we insert this value of t in the equation for ξ, we obtain

$$\xi = a\frac{c^2}{c^2 - v^2}x'.$$

In an analogous manner we find, by considering rays moving along the two other axes, that

$$\eta = c\tau = ac\left(t - \frac{v}{c^2 - v^2}x'\right)$$

when

$$\frac{y}{\sqrt{c^2 - v^2}} = t, \ x' = 0.$$

Thus

$$\eta = a\frac{c}{\sqrt{c^2 - v^2}}y \text{ and } \zeta = a\frac{c}{\sqrt{c^2 - v^2}}z.$$

Substituting for x' its value, we obtain

$$
\begin{aligned}
\tau &= \phi(v)\beta(t - vx/c^2), \\
\xi &= \phi(v)\beta(x - vt), \\
\eta &= \phi(v)y, \\
\zeta &= \phi(v)z,
\end{aligned}
$$

where

$$\beta = \frac{1}{\sqrt{1 - v^2/c^2}},$$

and ϕ is an as yet unknown function of v. If no assumption whatever be made as to the initial position of the moving system and as to the zero point of τ, an additive constant is to be placed on the right side of each of these equations.

We now have to prove that any ray of light, measured in the moving system, is propagated with the velocity c, if, as we have assumed, this is the case in the stationary system; for we have not as yet furnished the proof that the principle of the constancy of the velocity of light is compatible with the principle of relativity.

At the time $t = \tau = 0$, when the origin of the co-ordinates is common to the two systems, let a spherical wave be emitted therefrom, and be propagated with the velocity c in system K. If (x, y, z) be a point just attained by this wave, then

$$x^2 + y^2 + z^2 = c^2 t^2.$$

Transforming this equation with the aid of our equations of transformation we obtain after a simple calculation

$$\xi^2 + \eta^2 + \zeta^2 = c^2 \tau^2.$$

The wave under consideration is therefore no less a spherical wave with velocity of propagation c when viewed in the moving system. This shows that our two fundamental principles are compatible.[5]

In the equations of transformation which have been developed there enters an unknown function ϕ of v, which we will now determine.

For this purpose we introduce a third system of co-ordinates K', which relatively to the system k is in a state of parallel translatory motion parallel to the axis of Ξ,[†] such that the origin of co-ordinates of system K' moves with velocity $-v$ on the axis of Ξ. At the time $t = 0$ let all three origins coincide, and when $t = x = y = z = 0$ let the time t' of the system K' be zero. We call the co-ordinates, measured in the system K', x', y', z', and by a twofold application of our equations of transformation we obtain

$$
\begin{aligned}
t' &= \phi(-v)\beta(-v)(\tau + v\xi/c^2) &&= \phi(v)\phi(-v)t, \\
x' &= \phi(-v)\beta(-v)(\xi + v\tau) &&= \phi(v)\phi(-v)x, \\
y' &= \phi(-v)\eta &&= \phi(v)\phi(-v)y, \\
z' &= \phi(-v)\zeta &&= \phi(v)\phi(-v)z.
\end{aligned}
$$

Since the relations between x', y', z' and x, y, z do not contain the time t, the systems K and K' are at rest with respect to one another, and it is clear that the transformation from K to K' must be the identical transformation. Thus

$$\phi(v)\phi(-v) = 1.$$

---

[5]The equations of the Lorentz transformation may be more simply deduced directly from the condition that in virtue of those equations the relation $x^2 + y^2 + z^2 = c^2 t^2$ shall have as its consequence the second relation $\xi^2 + \eta^2 + \zeta^2 = c^2 \tau^2$.

[†]Editor's note: In Einstein's original paper, the symbols (Ξ, H, Z) for the co-ordinates of the moving system k were introduced without explicitly defining them. In the 1923 English translation, (X, Y, Z) were used, creating an ambiguity between X co-ordinates in the fixed system K and the parallel axis in moving system k. Here and in subsequent references we use Ξ when referring to the axis of system k along which the system is translating with respect to K. In addition, the reference to system K' later in this sentence was incorrectly given as "k" in the 1923 English translation.

We now inquire into the signification of $\phi(v)$. We give our attention to that part of the axis of Y of system k which lies between $\xi = 0, \eta = 0, \zeta = 0$ and $\xi = 0, \eta = l, \zeta = 0$. This part of the axis of Y is a rod moving perpendicularly to its axis with velocity v relatively to system K. Its ends possess in K the co-ordinates

$$x_1 = vt, \; y_1 = \frac{l}{\phi(v)}, \; z_1 = 0$$

and

$$x_2 = vt, \; y_2 = 0, \; z_2 = 0.$$

The length of the rod measured in K is therefore $l/\phi(v)$; and this gives us the meaning of the function $\phi(v)$. From reasons of symmetry it is now evident that the length of a given rod moving perpendicularly to its axis, measured in the stationary system, must depend only on the velocity and not on the direction and the sense of the motion. The length of the moving rod measured in the stationary system does not change, therefore, if v and $-v$ are interchanged. Hence follows that $l/\phi(v) = l/\phi(-v)$, or

$$\phi(v) = \phi(-v).$$

It follows from this relation and the one previously found that $\phi(v) = 1$, so that the transformation equations which have been found become

$$
\begin{aligned}
\tau &= \beta(t - vx/c^2), \\
\xi &= \beta(x - vt), \\
\eta &= y, \\
\zeta &= z,
\end{aligned}
$$

where

$$\beta = 1/\sqrt{1 - v^2/c^2}.$$

§ 4. Physical Meaning of the Equations Obtained in Respect to Moving Rigid Bodies and Moving Clocks

We envisage a rigid sphere[6] of radius R, at rest relatively to the moving system k, and with its centre at the origin of co-ordinates of k. The equation of the surface of this sphere moving relatively to the system K with velocity v is

$$\xi^2 + \eta^2 + \zeta^2 = R^2.$$

---

[6] That is, a body possessing spherical form when examined at rest.

The equation of this surface expressed in x, y, z at the time $t = 0$ is

$$\frac{x^2}{(\sqrt{1 - v^2/c^2})^2} + y^2 + z^2 = \mathrm{R}^2.$$

A rigid body which, measured in a state of rest, has the form of a sphere, therefore has in a state of motion—viewed from the stationary system—the form of an ellipsoid of revolution with the axes

$$\mathrm{R}\sqrt{1 - v^2/c^2}, \ \mathrm{R}, \ \mathrm{R}.$$

Thus, whereas the Y and Z dimensions of the sphere (and therefore of every rigid body of no matter what form) do not appear modified by the motion, the X dimension appears shortened in the ratio $1 : \sqrt{1 - v^2/c^2}$, i.e. the greater the value of v, the greater the shortening. For $v = c$ all moving objects—viewed from the "stationary" system—shrivel up into plane figures.[†] For velocities greater than that of light our deliberations become meaningless; we shall, however, find in what follows, that the velocity of light in our theory plays the part, physically, of an infinitely great velocity.

It is clear that the same results hold good of bodies at rest in the "stationary" system, viewed from a system in uniform motion.

Further, we imagine one of the clocks which are qualified to mark the time t when at rest relatively to the stationary system, and the time τ when at rest relatively to the moving system, to be located at the origin of the co-ordinates of k, and so adjusted that it marks the time τ. What is the rate of this clock, when viewed from the stationary system?

Between the quantities x, t, and τ, which refer to the position of the clock, we have, evidently, $x = vt$ and

$$\tau = \frac{1}{\sqrt{1 - v^2/c^2}}(t - vx/c^2).$$

Therefore,

$$\tau = t\sqrt{1 - v^2/c^2} = t - (1 - \sqrt{1 - v^2/c^2})t$$

whence it follows that the time marked by the clock (viewed in the stationary system) is slow by $1 - \sqrt{1 - v^2/c^2}$ seconds per second, or—neglecting magnitudes of fourth and higher order—by $\frac{1}{2}v^2/c^2$.

From this there ensues the following peculiar consequence. If at the points A and B of K there are stationary clocks which, viewed in the stationary system, are synchronous; and if the clock at A is moved with the velocity v along the line AB to B, then on its arrival at B the two clocks no longer synchronize, but the clock moved from A to B lags behind the other which has remained at

---

[†]Editor's note: In the 1923 English translation, this phrase was erroneously translated as "plain figures". I have used the correct "plane figures" in this edition.

B by $\frac{1}{2}tv^2/c^2$ (up to magnitudes of fourth and higher order), t being the time occupied in the journey from A to B.

It is at once apparent that this result still holds good if the clock moves from A to B in any polygonal line, and also when the points A and B coincide.

If we assume that the result proved for a polygonal line is also valid for a continuously curved line, we arrive at this result: If one of two synchronous clocks at A is moved in a closed curve with constant velocity until it returns to A, the journey lasting t seconds, then by the clock which has remained at rest the travelled clock on its arrival at A will be $\frac{1}{2}tv^2/c^2$ second slow. Thence we conclude that a balance-clock[7] at the equator must go more slowly, by a very small amount, than a precisely similar clock situated at one of the poles under otherwise identical conditions.

§ 5. The Composition of Velocities

In the system k moving along the axis of X of the system K with velocity v, let a point move in accordance with the equations

$$\xi = w_\xi \tau, \eta = w_\eta \tau, \zeta = 0,$$

where w_ξ and w_η denote constants.

Required: the motion of the point relatively to the system K. If with the help of the equations of transformation developed in § 3 we introduce the quantities x, y, z, t into the equations of motion of the point, we obtain

$$x = \frac{w_\xi + v}{1 + vw_\xi/c^2} t,$$

$$y = \frac{\sqrt{1 - v^2/c^2}}{1 + vw_\xi/c^2} w_\eta t,$$

$$z = 0.$$

Thus the law of the parallelogram of velocities is valid according to our theory only to a first approximation. We set

$$V^2 = \left(\frac{dx}{dt}\right)^2 + \left(\frac{dy}{dt}\right)^2,$$

$$w^2 = w_\xi^2 + w_\eta^2,$$

$$a = \tan^{-1} w_\eta/w_\xi,{}^\dagger$$

---

[7]Not a pendulum-clock, which is physically a system to which the Earth belongs. This case had to be excluded.

†Editor's note: This equation was incorrectly given in Einstein's original paper and the 1923 English translation as $a = \tan^{-1} w_y/w_x$.

a is then to be looked upon as the angle between the velocities v and w. After a simple calculation we obtain

$$V = \frac{\sqrt{(v^2 + w^2 + 2vw\cos a) - (vw\sin a/c)^2}}{1 + vw\cos a/c^2}.$$

It is worthy of remark that v and w enter into the expression for the resultant velocity in a symmetrical manner. If w also has the direction of the axis of X, we get

$$V = \frac{v + w}{1 + vw/c^2}.$$

It follows from this equation that from a composition of two velocities which are less than c, there always results a velocity less than c. For if we set $v = c - \kappa, w = c - \lambda$, κ and λ being positive and less than c, then

$$V = c\frac{2c - \kappa - \lambda}{2c - \kappa - \lambda + \kappa\lambda/c} < c.$$

It follows, further, that the velocity of light c cannot be altered by composition with a velocity less than that of light. For this case we obtain

$$V = \frac{c + w}{1 + w/c} = c.$$

We might also have obtained the formula for V, for the case when v and w have the same direction, by compounding two transformations in accordance with § 3. If in addition to the systems K and k figuring in § 3 we introduce still another system of co-ordinates k' moving parallel to k, its initial point moving on the axis of $\Xi^\dagger$ with the velocity w, we obtain equations between the quantities x, y, z, t and the corresponding quantities of k', which differ from the equations found in § 3 only in that the place of "v" is taken by the quantity

$$\frac{v + w}{1 + vw/c^2};$$

from which we see that such parallel transformations—necessarily—form a group.

We have now deduced the requisite laws of the theory of kinematics corresponding to our two principles, and we proceed to show their application to electrodynamics.

II. ELECTRODYNAMICAL PART

§ 6. Transformation of the Maxwell-Hertz Equations for Empty Space. On the Nature of the Electromotive Forces Occurring in a Magnetic Field During Motion

Let the Maxwell-Hertz equations for empty space hold good for the stationary system K, so that we have

---

†Editor's note: "X" in the 1923 English translation.

$$\frac{1}{c}\frac{\partial X}{\partial t} = \frac{\partial N}{\partial y} - \frac{\partial M}{\partial z}, \quad \frac{1}{c}\frac{\partial L}{\partial t} = \frac{\partial Y}{\partial z} - \frac{\partial Z}{\partial y},$$

$$\frac{1}{c}\frac{\partial Y}{\partial t} = \frac{\partial L}{\partial z} - \frac{\partial N}{\partial x}, \quad \frac{1}{c}\frac{\partial M}{\partial t} = \frac{\partial Z}{\partial x} - \frac{\partial X}{\partial z},$$

$$\frac{1}{c}\frac{\partial Z}{\partial t} = \frac{\partial M}{\partial x} - \frac{\partial L}{\partial y}, \quad \frac{1}{c}\frac{\partial N}{\partial t} = \frac{\partial X}{\partial y} - \frac{\partial Y}{\partial x},$$

where (X, Y, Z) denotes the vector of the electric force, and (L, M, N) that of the magnetic force.

If we apply to these equations the transformation developed in § 3, by referring the electromagnetic processes to the system of co-ordinates there introduced, moving with the velocity v, we obtain the equations

$$\frac{1}{c}\frac{\partial X}{\partial \tau} = \frac{\partial}{\partial \eta}\left\{\beta\left(N - \frac{v}{c}Y\right)\right\} - \frac{\partial}{\partial \zeta}\left\{\beta\left(M + \frac{v}{c}Z\right)\right\},$$

$$\frac{1}{c}\frac{\partial}{\partial \tau}\left\{\beta\left(Y - \frac{v}{c}N\right)\right\} = \frac{\partial L}{\partial \xi} - \frac{\partial}{\partial \zeta}\left\{\beta\left(N - \frac{v}{c}Y\right)\right\},$$

$$\frac{1}{c}\frac{\partial}{\partial \tau}\left\{\beta\left(Z + \frac{v}{c}M\right)\right\} = \frac{\partial}{\partial \xi}\left\{\beta\left(M + \frac{v}{c}Z\right)\right\} - \frac{\partial L}{\partial \eta},$$

$$\frac{1}{c}\frac{\partial L}{\partial \tau} = \frac{\partial}{\partial \zeta}\left\{\beta\left(Y - \frac{v}{c}N\right)\right\} - \frac{\partial}{\partial \eta}\left\{\beta\left(Z + \frac{v}{c}M\right)\right\},$$

$$\frac{1}{c}\frac{\partial}{\partial \tau}\left\{\beta\left(M + \frac{v}{c}Z\right)\right\} = \frac{\partial}{\partial \xi}\left\{\beta\left(Z + \frac{v}{c}M\right)\right\} - \frac{\partial X}{\partial \zeta},$$

$$\frac{1}{c}\frac{\partial}{\partial \tau}\left\{\beta\left(N - \frac{v}{c}Y\right)\right\} = \frac{\partial X}{\partial \eta} - \frac{\partial}{\partial \xi}\left\{\beta\left(Y - \frac{v}{c}N\right)\right\},$$

where

$$\beta = 1/\sqrt{1 - v^2/c^2}.$$

Now the principle of relativity requires that if the Maxwell-Hertz equations for empty space hold good in system K, they also hold good in system k; that is to say that the vectors of the electric and the magnetic force—(X′, Y′, Z′) and (L′, M′, N′)—of the moving system k, which are defined by their ponderomotive effects on electric or magnetic masses respectively, satisfy the following equations:—

$$\frac{1}{c}\frac{\partial X'}{\partial \tau} = \frac{\partial N'}{\partial \eta} - \frac{\partial M'}{\partial \zeta}, \quad \frac{1}{c}\frac{\partial L'}{\partial \tau} = \frac{\partial Y'}{\partial \zeta} - \frac{\partial Z'}{\partial \eta},$$

$$\frac{1}{c}\frac{\partial Y'}{\partial \tau} = \frac{\partial L'}{\partial \zeta} - \frac{\partial N'}{\partial \xi}, \quad \frac{1}{c}\frac{\partial M'}{\partial \tau} = \frac{\partial Z'}{\partial \xi} - \frac{\partial X'}{\partial \zeta},$$

$$\frac{1}{c}\frac{\partial Z'}{\partial \tau} = \frac{\partial M'}{\partial \xi} - \frac{\partial L'}{\partial \eta}, \quad \frac{1}{c}\frac{\partial N'}{\partial \tau} = \frac{\partial X'}{\partial \eta} - \frac{\partial Y'}{\partial \xi}.$$

Evidently the two systems of equations found for system k must express exactly the same thing, since both systems of equations are equivalent to the Maxwell-Hertz equations for system K. Since, further, the equations of the two systems agree, with the exception of the symbols for the vectors, it follows that the functions occurring in the systems of equations at corresponding places must agree, with the exception of a factor $\psi(v)$, which is common for all functions of the one system of equations, and is independent of ξ, η, ζ and τ but depends upon v. Thus we have the relations

$$
\begin{aligned}
X' &= \psi(v)X, & L' &= \psi(v)L, \\
Y' &= \psi(v)\beta\left(Y - \tfrac{v}{c}N\right), & M' &= \psi(v)\beta\left(M + \tfrac{v}{c}Z\right), \\
Z' &= \psi(v)\beta\left(Z + \tfrac{v}{c}M\right), & N' &= \psi(v)\beta\left(N - \tfrac{v}{c}Y\right).
\end{aligned}
$$

If we now form the reciprocal of this system of equations, firstly by solving the equations just obtained, and secondly by applying the equations to the inverse transformation (from k to K), which is characterized by the velocity $-v$, it follows, when we consider that the two systems of equations thus obtained must be identical, that $\psi(v)\psi(-v) = 1$. Further, from reasons of symmetry[8] and therefore

$$
\psi(v) = 1,
$$

and our equations assume the form

$$
\begin{aligned}
X' &= X, & L' &= L, \\
Y' &= \beta\left(Y - \tfrac{v}{c}N\right), & M' &= \beta\left(M + \tfrac{v}{c}Z\right), \\
Z' &= \beta\left(Z + \tfrac{v}{c}M\right), & N' &= \beta\left(N - \tfrac{v}{c}Y\right).
\end{aligned}
$$

As to the interpretation of these equations we make the following remarks: Let a point charge of electricity have the magnitude "one" when measured in the stationary system K, i.e. let it when at rest in the stationary system exert a force of one dyne upon an equal quantity of electricity at a distance of one cm. By the principle of relativity this electric charge is also of the magnitude "one" when measured in the moving system. If this quantity of electricity is at rest relatively to the stationary system, then by definition the vector (X, Y, Z) is equal to the force acting upon it. If the quantity of electricity is at rest relatively to the moving system (at least at the relevant instant), then the force acting upon it, measured in the moving system, is equal to the vector (X', Y', Z'). Consequently the first three equations above allow themselves to be clothed in words in the two following ways:—

1. If a unit electric point charge is in motion in an electromagnetic field, there acts upon it, in addition to the electric force, an "electromotive force" which, if we neglect the terms multiplied by the second and higher powers of v/c, is equal to the vector-product of the velocity of the charge and the magnetic force, divided by the velocity of light. (Old manner of expression.)

---

[8]If, for example, X=Y=Z=L=M=0, and N $\neq$ 0, then from reasons of symmetry it is clear that when v changes sign without changing its numerical value, Y' must also change sign without changing its numerical value.

2. If a unit electric point charge is in motion in an electromagnetic field, the force acting upon it is equal to the electric force which is present at the locality of the charge, and which we ascertain by transformation of the field to a system of co-ordinates at rest relatively to the electrical charge. (New manner of expression.)

The analogy holds with "magnetomotive forces." We see that electromotive force plays in the developed theory merely the part of an auxiliary concept, which owes its introduction to the circumstance that electric and magnetic forces do not exist independently of the state of motion of the system of co-ordinates.

Furthermore it is clear that the asymmetry mentioned in the introduction as arising when we consider the currents produced by the relative motion of a magnet and a conductor, now disappears. Moreover, questions as to the "seat" of electrodynamic electromotive forces (unipolar machines) now have no point.

§ 7. Theory of Doppler's Principle and of Aberration

In the system K, very far from the origin of co-ordinates, let there be a source of electrodynamic waves, which in a part of space containing the origin of co-ordinates may be represented to a sufficient degree of approximation by the equations

$$
\begin{aligned}
X &= X_0 \sin \Phi, \quad L = L_0 \sin \Phi, \\
Y &= Y_0 \sin \Phi, \quad M = M_0 \sin \Phi, \\
Z &= Z_0 \sin \Phi, \quad N = N_0 \sin \Phi,
\end{aligned}
$$

where

$$
\Phi = \omega \left\{ t - \frac{1}{c}(lx + my + nz) \right\}.
$$

Here (X_0, Y_0, Z_0) and (L_0, M_0, N_0) are the vectors defining the amplitude of the wave-train, and l, m, n the direction-cosines of the wave-normals. We wish to know the constitution of these waves, when they are examined by an observer at rest in the moving system k.

Applying the equations of transformation found in § 6 for electric and magnetic forces, and those found in § 3 for the co-ordinates and the time, we obtain directly

$$
\begin{aligned}
X' &= X_0 \sin \Phi', \qquad\qquad L' = L_0 \sin \Phi', \\
Y' &= \beta(Y_0 - vN_0/c) \sin \Phi', \quad M' = \beta(M_0 + vZ_0/c) \sin \Phi', \\
Z' &= \beta(Z_0 + vM_0/c) \sin \Phi', \quad N' = \beta(N_0 - vY_0/c) \sin \Phi', \\
\Phi' &= \omega' \left\{ \tau - \tfrac{1}{c}(l'\xi + m'\eta + n'\zeta) \right\}
\end{aligned}
$$

where

$$
\omega' \;=\; \omega\beta(1 - lv/c),
$$

$$l' = \frac{l - v/c}{1 - lv/c},$$

$$m' = \frac{m}{\beta(1 - lv/c)},$$

$$n' = \frac{n}{\beta(1 - lv/c)}.$$

From the equation for ω' it follows that if an observer is moving with velocity v relatively to an infinitely distant source of light of frequency ν, in such a way that the connecting line "source-observer" makes the angle ϕ with the velocity of the observer referred to a system of co-ordinates which is at rest relatively to the source of light, the frequency ν' of the light perceived by the observer is given by the equation

$$\nu' = \nu \frac{1 - \cos \phi \cdot v/c}{\sqrt{1 - v^2/c^2}}.$$

This is Doppler's principle for any velocities whatever. When $\phi = 0$ the equation assumes the perspicuous form

$$\nu' = \nu \sqrt{\frac{1 - v/c}{1 + v/c}}.$$

We see that, in contrast with the customary view, when $v = -c, \nu' = \infty$.

If we call the angle between the wave-normal (direction of the ray) in the moving system and the connecting line "source-observer" ϕ', the equation for $\phi'^\dagger$ assumes the form

$$\cos \phi' = \frac{\cos \phi - v/c}{1 - \cos \phi \cdot v/c}.$$

This equation expresses the law of aberration in its most general form. If $\phi = \frac{1}{2}\pi$, the equation becomes simply

$$\cos \phi' = -v/c.$$

We still have to find the amplitude of the waves, as it appears in the moving system. If we call the amplitude of the electric or magnetic force A or A' respectively, accordingly as it is measured in the stationary system or in the moving system, we obtain

$$A'^2 = A^2 \frac{(1 - \cos \phi \cdot v/c)^2}{1 - v^2/c^2}$$

which equation, if $\phi = 0$, simplifies into

---

†Editor's note: Erroneously given as "l'" in the 1923 English translation, propagating an error, despite a change in symbols, from the original 1905 paper.

$$A'^2 = A^2 \frac{1 - v/c}{1 + v/c}.$$

It follows from these results that to an observer approaching a source of light with the velocity c, this source of light must appear of infinite intensity.

§ 8. Transformation of the Energy of Light Rays. Theory of the Pressure of Radiation Exerted on Perfect Reflectors

Since $A^2/8\pi$ equals the energy of light per unit of volume, we have to regard $A'^2/8\pi$, by the principle of relativity, as the energy of light in the moving system. Thus A'^2/A^2 would be the ratio of the "measured in motion" to the "measured at rest" energy of a given light complex, if the volume of a light complex were the same, whether measured in K or in k. But this is not the case. If l, m, n are the direction-cosines of the wave-normals of the light in the stationary system, no energy passes through the surface elements of a spherical surface moving with the velocity of light:—

$$(x - lct)^2 + (y - mct)^2 + (z - nct)^2 = R^2.$$

We may therefore say that this surface permanently encloses the same light complex. We inquire as to the quantity of energy enclosed by this surface, viewed in system k, that is, as to the energy of the light complex relatively to the system k.

The spherical surface—viewed in the moving system—is an ellipsoidal surface, the equation for which, at the time $\tau = 0$, is

$$(\beta\xi - l\beta\xi v/c)^2 + (\eta - m\beta\xi v/c)^2 + (\zeta - n\beta\xi v/c)^2 = R^2.$$

If S is the volume of the sphere, and S′ that of this ellipsoid, then by a simple calculation

$$\frac{S'}{S} = \frac{\sqrt{1 - v^2/c^2}}{1 - \cos\phi \cdot v/c}.$$

Thus, if we call the light energy enclosed by this surface E when it is measured in the stationary system, and E′ when measured in the moving system, we obtain

$$\frac{E'}{E} = \frac{A'^2 S'}{A^2 S} = \frac{1 - \cos\phi \cdot v/c}{\sqrt{1 - v^2/c^2}},$$

and this formula, when $\phi = 0$, simplifies into

$$\frac{E'}{E} = \sqrt{\frac{1 - v/c}{1 + v/c}}.$$

It is remarkable that the energy and the frequency of a light complex vary with the state of motion of the observer in accordance with the same law.

Now let the co-ordinate plane $\xi = 0$ be a perfectly reflecting surface, at which the plane waves considered in § 7 are reflected. We seek for the pressure of light exerted on the reflecting surface, and for the direction, frequency, and intensity of the light after reflexion.

Let the incidental light be defined by the quantities A, $\cos\phi$, ν (referred to system K). Viewed from k the corresponding quantities are

$$
\begin{aligned}
A' &= A\frac{1 - \cos\phi \cdot v/c}{\sqrt{1 - v^2/c^2}}, \\
\cos\phi' &= \frac{\cos\phi - v/c}{1 - \cos\phi \cdot v/c}, \\
\nu' &= \nu\frac{1 - \cos\phi \cdot v/c}{\sqrt{1 - v^2/c^2}}.
\end{aligned}
$$

For the reflected light, referring the process to system k, we obtain

$$
\begin{aligned}
A'' &= A' \\
\cos\phi'' &= -\cos\phi' \\
\nu'' &= \nu'
\end{aligned}
$$

Finally, by transforming back to the stationary system K, we obtain for the reflected light

$$
\begin{aligned}
A''' &= A''\frac{1 + cos\phi'' \cdot v/c}{\sqrt{1 - v^2/c^2}} = A\frac{1 - 2\cos\phi \cdot v/c + v^2/c^2}{1 - v^2/c^2}, \\
\cos\phi''' &= \frac{\cos\phi'' + v/c}{1 + \cos\phi'' \cdot v/c} = -\frac{(1 + v^2/c^2)\cos\phi - 2v/c}{1 - 2\cos\phi \cdot v/c + v^2/c^2}, \\
\nu''' &= \nu''\frac{1 + \cos\phi'' \cdot v/c}{\sqrt{1 - v^2/c^2}} = \nu\frac{1 - 2\cos\phi \cdot v/c + v^2/c^2}{1 - v^2/c^2}.
\end{aligned}
$$

The energy (measured in the stationary system) which is incident upon unit area of the mirror in unit time is evidently $A^2(c\cos\phi - v)/8\pi$. The energy leaving the unit of surface of the mirror in the unit of time is $A'''^2(-c\cos\phi''' + v)/8\pi$. The difference of these two expressions is, by the principle of energy, the work done by the pressure of light in the unit of time. If we set down this work as equal to the product Pv, where P is the pressure of light, we obtain

$$
P = 2 \cdot \frac{A^2}{8\pi}\frac{(\cos\phi - v/c)^2}{1 - v^2/c^2}.
$$

In agreement with experiment and with other theories, we obtain to a first approximation

$$P = 2 \cdot \frac{A^2}{8\pi} \cos^2 \phi.$$

All problems in the optics of moving bodies can be solved by the method here employed. What is essential is, that the electric and magnetic force of the light which is influenced by a moving body, be transformed into a system of co-ordinates at rest relatively to the body. By this means all problems in the optics of moving bodies will be reduced to a series of problems in the optics of stationary bodies.

§ 9. Transformation of the Maxwell-Hertz Equations when Convection-Currents are Taken into Account

We start from the equations

$$\frac{1}{c}\left\{\frac{\partial X}{\partial t} + u_x\rho\right\} = \frac{\partial N}{\partial y} - \frac{\partial M}{\partial z}, \quad \frac{1}{c}\frac{\partial L}{\partial t} = \frac{\partial Y}{\partial z} - \frac{\partial Z}{\partial y},$$

$$\frac{1}{c}\left\{\frac{\partial Y}{\partial t} + u_y\rho\right\} = \frac{\partial L}{\partial z} - \frac{\partial N}{\partial x}, \quad \frac{1}{c}\frac{\partial M}{\partial t} = \frac{\partial Z}{\partial x} - \frac{\partial X}{\partial z},$$

$$\frac{1}{c}\left\{\frac{\partial Z}{\partial t} + u_z\rho\right\} = \frac{\partial M}{\partial x} - \frac{\partial L}{\partial y}, \quad \frac{1}{c}\frac{\partial N}{\partial t} = \frac{\partial X}{\partial y} - \frac{\partial Y}{\partial x},$$

where

$$\rho = \frac{\partial X}{\partial x} + \frac{\partial Y}{\partial y} + \frac{\partial Z}{\partial z}$$

denotes 4π times the density of electricity, and (u_x, u_y, u_z) the velocity-vector of the charge. If we imagine the electric charges to be invariably coupled to small rigid bodies (ions, electrons), these equations are the electromagnetic basis of the Lorentzian electrodynamics and optics of moving bodies.

Let these equations be valid in the system K, and transform them, with the assistance of the equations of transformation given in §§ 3 and 6, to the system k. We then obtain the equations

$$\frac{1}{c}\left\{\frac{\partial X'}{\partial \tau} + u_\xi\rho'\right\} = \frac{\partial N'}{\partial \eta} - \frac{\partial M'}{\partial \zeta}, \quad \frac{1}{c}\frac{\partial L'}{\partial \tau} = \frac{\partial Y'}{\partial \zeta} - \frac{\partial Z'}{\partial \eta},$$

$$\frac{1}{c}\left\{\frac{\partial Y'}{\partial \tau} + u_\eta\rho'\right\} = \frac{\partial L'}{\partial \zeta} - \frac{\partial N'}{\partial \xi}, \quad \frac{1}{c}\frac{\partial M'}{\partial \tau} = \frac{\partial Z'}{\partial \xi} - \frac{\partial X'}{\partial \zeta},$$

$$\frac{1}{c}\left\{\frac{\partial Z'}{\partial \tau} + u_\zeta\rho'\right\} = \frac{\partial M'}{\partial \xi} - \frac{\partial L'}{\partial \eta}, \quad \frac{1}{c}\frac{\partial N'}{\partial \tau} = \frac{\partial X'}{\partial \eta} - \frac{\partial Y'}{\partial \xi},$$

where

$$u_\xi = \frac{u_x - v}{1 - u_x v/c^2}$$

$$u_\eta = \frac{u_y}{\beta(1 - u_x v/c^2)}$$

$$u_\zeta = \frac{u_z}{\beta(1 - u_x v/c^2)},$$

and

$$\rho' = \frac{\partial X'}{\partial \xi} + \frac{\partial Y'}{\partial \eta} + \frac{\partial Z'}{\partial \zeta}$$

$$= \beta(1 - u_x v/c^2)\rho.$$

Since—as follows from the theorem of addition of velocities (§ 5)—the vector (u_ξ, u_η, u_ζ) is nothing else than the velocity of the electric charge, measured in the system k, we have the proof that, on the basis of our kinematical principles, the electrodynamic foundation of Lorentz's theory of the electrodynamics of moving bodies is in agreement with the principle of relativity.

In addition I may briefly remark that the following important law may easily be deduced from the developed equations: If an electrically charged body is in motion anywhere in space without altering its charge when regarded from a system of co-ordinates moving with the body, its charge also remains—when regarded from the "stationary" system K—constant.

§ 10. Dynamics of the Slowly Accelerated Electron

Let there be in motion in an electromagnetic field an electrically charged particle (in the sequel called an "electron"), for the law of motion of which we assume as follows:—

If the electron is at rest at a given epoch, the motion of the electron ensues in the next instant of time according to the equations

$$m\frac{d^2x}{dt^2} = \epsilon X$$

$$m\frac{d^2y}{dt^2} = \epsilon Y$$

$$m\frac{d^2z}{dt^2} = \epsilon Z$$

where x, y, z denote the co-ordinates of the electron, and m the mass of the electron, as long as its motion is slow.

Now, secondly, let the velocity of the electron at a given epoch be v. We seek the law of motion of the electron in the immediately ensuing instants of time.

Without affecting the general character of our considerations, we may and will assume that the electron, at the moment when we give it our attention, is at the origin of the co-ordinates, and moves with the velocity v along the axis of X of the system K. It is then clear that at the given moment ($t = 0$) the electron is at rest relatively to a system of co-ordinates which is in parallel motion with velocity v along the axis of X.

From the above assumption, in combination with the principle of relativity, it is clear that in the immediately ensuing time (for small values of t) the electron, viewed from the system k, moves in accordance with the equations

$$m\frac{d^2\xi}{d\tau^2} = \epsilon X',$$

$$m\frac{d^2\eta}{d\tau^2} = \epsilon Y',$$

$$m\frac{d^2\zeta}{d\tau^2} = \epsilon Z',$$

in which the symbols ξ, η, ζ, X', Y', Z' refer to the system k. If, further, we decide that when $t = x = y = z = 0$ then $\tau = \xi = \eta = \zeta = 0$, the transformation equations of §§ 3 and 6 hold good, so that we have

$$\xi = \beta(x - vt), \eta = y, \zeta = z, \tau = \beta(t - vx/c^2),$$
$$X' = X, Y' = \beta(Y - vN/c), Z' = \beta(Z + vM/c).$$

With the help of these equations we transform the above equations of motion from system k to system K, and obtain

$$\left.\begin{array}{rcl} \frac{d^2x}{dt^2} &=& \frac{\epsilon}{m\beta^3}X \\ \frac{d^2y}{dt^2} &=& \frac{\epsilon}{m\beta}\left(Y - \frac{v}{c}N\right) \\ \frac{d^2z}{dt^2} &=& \frac{\epsilon}{m\beta}\left(Z + \frac{v}{c}M\right) \end{array}\right\} \quad \cdot \quad \cdot \quad \cdot \quad (A)$$

Taking the ordinary point of view we now inquire as to the "longitudinal" and the "transverse" mass of the moving electron. We write the equations (A) in the form

$$\begin{array}{rclcl} m\beta^3\frac{d^2x}{dt^2} &=& \epsilon X &=& \epsilon X', \\ m\beta^2\frac{d^2y}{dt^2} &=& \epsilon\beta\left(Y - \frac{v}{c}N\right) &=& \epsilon Y', \\ m\beta^2\frac{d^2z}{dt^2} &=& \epsilon\beta\left(Z + \frac{v}{c}M\right) &=& \epsilon Z', \end{array}$$

and remark firstly that $\epsilon X'$, $\epsilon Y'$, $\epsilon Z'$ are the components of the ponderomotive force acting upon the electron, and are so indeed as viewed in a system moving at the moment with the electron, with the same velocity as the electron. (This force might be measured, for example, by a spring balance at rest in the last-mentioned system.) Now if we call this force simply "the force acting upon the electron,"[9] and maintain the equation—mass × acceleration = force—and if we also decide that the accelerations are to be measured in the stationary system K, we derive from the above equations

$$\text{Longitudinal mass} \quad = \quad \frac{m}{(\sqrt{1 - v^2/c^2})^3}.$$

$$\text{Transverse mass} \quad = \quad \frac{m}{1 - v^2/c^2}.$$

With a different definition of force and acceleration we should naturally obtain other values for the masses. This shows us that in comparing different theories of the motion of the electron we must proceed very cautiously.

We remark that these results as to the mass are also valid for ponderable material points, because a ponderable material point can be made into an electron (in our sense of the word) by the addition of an electric charge, *no matter how small.*

We will now determine the kinetic energy of the electron. If an electron moves from rest at the origin of co-ordinates of the system K along the axis of X under the action of an electrostatic force X, it is clear that the energy withdrawn from the electrostatic field has the value $\int \epsilon X \, dx$. As the electron is to be slowly accelerated, and consequently may not give off any energy in the form of radiation, the energy withdrawn from the electrostatic field must be put down as equal to the energy of motion W of the electron. Bearing in mind that during the whole process of motion which we are considering, the first of the equations (A) applies, we therefore obtain

$$\begin{aligned} \text{W} \quad &= \quad \int \epsilon X \, dx = m \int_0^v \beta^3 v \, dv \\ &= \quad mc^2 \left\{ \frac{1}{\sqrt{1 - v^2/c^2}} - 1 \right\}. \end{aligned}$$

Thus, when $v = c$, W becomes infinite. Velocities greater than that of light have—as in our previous results—no possibility of existence.

This expression for the kinetic energy must also, by virtue of the argument stated above, apply to ponderable masses as well.

We will now enumerate the properties of the motion of the electron which result from the system of equations (A), and are accessible to experiment.

1. From the second equation of the system (A) it follows that an electric force Y and a magnetic force N have an equally strong deflective action on an electron moving with the velocity v, when Y = Nv/c. Thus we see that it is possible by our theory to determine the velocity of the electron from the ratio

---

[9]The definition of force here given is not advantageous, as was first shown by M. Planck. It is more to the point to define force in such a way that the laws of momentum and energy assume the simplest form.

of the magnetic power of deflexion A_m to the electric power of deflexion A_e, for any velocity, by applying the law

$$\frac{A_m}{A_e} = \frac{v}{c}.$$

This relationship may be tested experimentally, since the velocity of the electron can be directly measured, e.g. by means of rapidly oscillating electric and magnetic fields.

2. From the deduction for the kinetic energy of the electron it follows that between the potential difference, P, traversed and the acquired velocity v of the electron there must be the relationship

$$P = \int X dx = \frac{m}{\epsilon} c^2 \left\{ \frac{1}{\sqrt{1 - v^2/c^2}} - 1 \right\}.$$

3. We calculate the radius of curvature of the path of the electron when a magnetic force N is present (as the only deflective force), acting perpendicularly to the velocity of the electron. From the second of the equations (A) we obtain

$$-\frac{d^2y}{dt^2} = \frac{v^2}{R} = \frac{\epsilon}{m} \frac{v}{c} N \sqrt{1 - \frac{v^2}{c^2}}$$

or

$$R = \frac{mc^2}{\epsilon} \cdot \frac{v/c}{\sqrt{1 - v^2/c^2}} \cdot \frac{1}{N}.$$

These three relationships are a complete expression for the laws according to which, by the theory here advanced, the electron must move.

In conclusion I wish to say that in working at the problem here dealt with I have had the loyal assistance of my friend and colleague M. Besso, and that I am indebted to him for several valuable suggestions.

This edition of Einstein's *On the Electrodynamics of Moving Bodies* is based on the English translation of his original 1905 German-language paper (published as *Zur Elektrodynamik bewegter Körper*, in *Annalen der Physik.* **17**:891, 1905) which appeared in the book *The Principle of Relativity*, published in 1923 by Methuen and Company, Ltd. of London. Most of the papers in that collection are English translations from the German *Das Relativatsprinzip*, 4th ed., published by in 1922 by Tuebner. All of these sources are now in the public domain; this document, derived from them, remains in the public domain and may be reproduced in any manner or medium without permission, restriction, attribution, or compensation.

Numbered footnotes are as they appeared in the 1923 edition; editor's notes are marked by a dagger (†) and appear in sans serif type. The 1923 English translation modified the notation used in Einstein's 1905 paper to conform to that in use by the 1920's; for example, c denotes the speed of light, as opposed the V used by Einstein in 1905.

This edition was prepared by John Walker. The current version of this document is available in a variety of formats from the editor's Web site:

http://www.fourmilab.ch/

Einstein's Papers on the General Theory of Relativity

"The Field Equations of Gravitation"
A. Einstein
November 25, 1915

The paper can be found online at:

In German
"Die Feldgleichungen der Gravitation,"
Preussische Akademie der Wissenschaften, Sitzungsberichte, Nov. 25, 1915 (part 2), 844–847, as accessed on the Internet in January 2015, go to:

https://ia601403.us.archive.org/25/items/sitzungsberichte1915deut/sitzungsberichte1915deut.pdf

In English
"The Field Equations of Gravitation" (1915)
by Albert Einstein, translated from German by Wikisource, as accessed on February 23, 2015, go to:

https://en.wikisource.org/?curid=735695

The Field Equations of Gravitation

By A. Einstein

I have shown in two recently published reports,[1] how one can arrive at field equations of gravitation, that are in agreement with the postulate of general relativity, *i.e.* which in their general form are covariant in respect to arbitrary substitutions of space-time variables.

The line of development was as follows. At first I found equations, that contain Newton's theory as approximation and that are covariant in respect to arbitrary substitutions of the determinant 1. Afterwards I found, that those equations in general correspond to covariant ones, if the scalar of the energy tensor of "matter" vanishes. The coordinate system had to be specialized in accordance with the simple rule, that $\sqrt{-g}$ is made to 1, whereby the equations of the theory experience an eminent simplification. In the course of this, however, one had to introduce the hypothesis, that the scalar of the energy tensor of matter vanishes.

Recently I find now, that one is able to dispense with hypothesis concerning the energy tensor of matter, if one fills in the energy tensor of matter into the field equations in a somehow different way than it was done in my two earlier reports. The field equations for vacuum, upon which I based the explanation of the perihelion motion of mercury, remain untouched by this modification. I give the complete consideration again at this place, so that the reader is not forced to uninterruptedly consult the earlier reports.

From the well known Riemannian covariant of fourth rank, the following covariant of second rank is derived:

$$G_{im} = R_{im} + S_{im} \tag{1}$$

$$R_{im} = -\sum_l \frac{\partial \left\{ \begin{smallmatrix} im \\ l \end{smallmatrix} \right\}}{\partial x_l} + \sum_{l\rho} \left\{ \begin{smallmatrix} il \\ \rho \end{smallmatrix} \right\} \left\{ \begin{smallmatrix} m\rho \\ l \end{smallmatrix} \right\} \tag{1a}$$

$$S_{im} = \sum_l \frac{\partial \left\{ \begin{smallmatrix} il \\ l \end{smallmatrix} \right\}}{\partial x_m} - \sum_{l\rho} \left\{ \begin{smallmatrix} im \\ \rho \end{smallmatrix} \right\} \left\{ \begin{smallmatrix} \rho l \\ l \end{smallmatrix} \right\} \tag{1b}$$

[845] We obtain the ten general covariant equations of the gravitational field in spaces, in which "matter" is absent, by putting

$$G_{im} = 0 \qquad (2)$$

These equations can be formed in a simpler way, when one choses the reference system so that $\sqrt{-g} = 1$. Then S_{im} vanishes due to (1b), so that one obtains instead of (2)

$$R_{im} = \sum_l \frac{\partial \Gamma^l_{im}}{\partial x_l} + \sum_{\rho l} \Gamma^l_{i\rho} \Gamma^\rho_{ml} = 0 \qquad (3)$$

$$\sqrt{-g} = 1 \qquad (3a)$$

Here we put

$$\Gamma^l_{im} = - \left\{ \begin{matrix} im \\ l \end{matrix} \right\} \qquad (4)$$

which magnitudes we will denote as the "components" of the gravitational field.

If "matter" exists in the considered space, then its energy tensor appears on the right hand side of (2) or (3). We put

$$G_{im} = -\varkappa \left(T_{im} - \frac{1}{2} g_{im} T \right) \qquad (2a)$$

where we put

$$\sum_{\rho\sigma} g^{\rho\sigma} T_{\rho\sigma} = \sum_\sigma T^\sigma_\sigma = T \qquad (5)$$

T is the scalar of the energy tensor of "matter", the right hand side of (2a) is a tensor. If we specialize the coordinate system in the ordinary way again, then we obtain instead of (2a) the equivalent equations

$$R_{im} = \sum_l \frac{\partial \Gamma^l_{im}}{\partial x_l} + \sum_{\rho l} \Gamma^l_{i\rho} \Gamma^\rho_{ml} = -\varkappa \left(T_{im} - \frac{1}{2} g_{im} T \right) \quad (6)$$

$$\sqrt{-g} = 1 \qquad (3a)$$

Like always we assume, that the divergence of the energy tensor of matter vanishes in the sense of the general differential calculus (Momentum-Energy theorem). When specializing the coordinate choice in accordance with (3a), it follows from it, that the T_{im} shall fulfill the conditions

$$\sum_\lambda \frac{\partial T^\lambda_\sigma}{\partial x_\lambda} = -\frac{1}{2} \sum_{\mu\nu} \frac{\partial g^{\mu\nu}}{\partial x_\sigma} T_{\mu\nu} \qquad (7)$$

or

$$\sum_\lambda \frac{\partial T_\sigma^\lambda}{\partial x_\lambda} = -\sum_{\mu\nu} \Gamma_{\sigma\nu}^\mu T_\mu^\nu \qquad (7a)$$

[846]

If one multiplies (6) by $\frac{\partial g^{im}}{\partial x_\sigma}$ and sums over i and m, then one obtains[2] in respect to (7) and in respect to the relation following from (3a)

$$\frac{1}{2} \sum_{im} g_{im} \frac{\partial g^{im}}{\partial x_\sigma} = -\frac{\partial lg\sqrt{-g}}{\partial x_\sigma} = 0$$

the conservation law for matter and the gravitational field together in the form

$$\sum_\lambda \frac{\partial}{\partial x_\lambda} \left(T_\sigma^\lambda + t_\sigma^\lambda \right) = 0 \qquad (8)$$

where t_σ^λ (the "energy tensor" of the gravitational field) is given by

$$\varkappa t_\sigma^\lambda = \frac{1}{2} \delta_\sigma^\lambda \sum_{\mu\nu\alpha\beta} g^{\mu\nu} \Gamma_{\mu\beta}^\alpha \Gamma_{\nu\alpha}^\beta - \sum g^{\mu\nu} \Gamma_{\mu\sigma}^\alpha \Gamma_{\nu\alpha}^\lambda \qquad (8a)$$

The reasons that drove me to the introduction of the second member on the right-hand side of (2a) and (6), become clear from the following considerations, that are completely analogous to those given at the place just mentioned (p. 785).

If we multiply (6) by g^{im} and sum over the indices i and m, then we obtain after simple calculation

$$\sum_{\alpha\beta} \frac{\partial^2 g^{\alpha\beta}}{\partial x_\alpha \partial x_\beta} - \varkappa(T + t) = 0 \qquad (9)$$

where corresponding to (5) it is put for abbreviation

$$\sum_{\rho\sigma} g^{\rho\sigma} t_{\rho\sigma} = \sum_\sigma t_\sigma^\sigma = t \qquad (8b)$$

Note, that it follows from the additional term, that in (9) the energy tensor of the gravitational field occurs besides that of matter in the same way, which is not the case in equations (21) l.c..

Furthermore one derives instead of equation (22) l.c., in the way as it is given there by the aid of the energy equation, the relations:

$$\frac{\partial}{\partial x_\mu} \left[\sum_{\alpha\beta} \frac{\partial^2 g^{\alpha\beta}}{\partial x_\alpha \partial x_\beta} - \varkappa(T + t) \right] = 0 \qquad (10)$$

[847]

From our additional term it follows, that these equations contain no new condition in respect to (9), so that concerning the energy tensor of matter, no other presupposition has to be made than the one, that it has to be in agreement with the momentum-energy theorem.

By that, the general theory of relativity as a logical building is eventually finished. The relativity postulate in its general form that makes the space-time coordinates to physically meaningless parameters, is directed with stringent necessity to a very specific theory of gravitation that explains the perihelion motion of mercury. However, the general relativity postulate offers nothing new about the essence of the other natural processes, which wasn't already taught by the special theory of relativity. My opinion regarding this issue, recently expressed at this place, was erroneous. Any physical theory equivalent to the special theory of relativity, can be included in the general theory of relativity by means of the absolute differential calculus, without that the latter gives any criterion for the admissibility of that theory.

1. ↑ Sitzungsber. XLIV, p. 778 and XLVI, p. 799, 1915
2. ↑ Concerning the derivation see Sitzungsber. XLIV, 1915, p. 784/785. For the following, I request the reader to use the derivations given on p. 785 for comparison.

The Foundation of the Generalized Theory of Relativity (1916)

Original German
as accessed in Janurary, 2015, go to:
http://www.physik.uni-augsburg.de/annalen/history/einstein-papers/1916_49_769-822.pdf

1920 English translation
by M.N. Saha and S.N. Bose, as accessed in Jan 2015, go to:
https://archive.org/details/principleofrelat00eins

Outline

1.   1 A. Principal considerations about the Postulate of Relativity.
     1.   1.1 § 1. Remarks on the Special Relativity Theory.
     2.   1.2 § 2. About the reasons which explain the extension of the relativity-postulate.
     3.   1.3 § 3. The time-space continuum. Requirements of the general Co-variance for the equations expressing the laws of Nature in general.
     4.   1.4 § 4. Relation of four co-ordinates to spatial and temporal measurements. Analytical expression for the Gravitation-field.
2.   2 B. Mathematical Auxiliaries for Establishing the General Covariant Equations.
     1.   2.1 § 5. Contravariant and covariant Four-vector.
     2.   2.2 § 6. Tensors of the second and higher ranks.
     3.   2.3 § 7. Multiplication of Tensors.
     4.   2.4 § 8. A few words about the Fundamental Tensor .
     5.   2.5 § 9. Equation of the geodetic line (or of point-motion).
     6.   2.6 § 10. Formation of Tensors through Differentiation.
     7.   2.7 § 11. Some special cases of Particular Importance.
     8.   2.8 § 12. The Riemann-Christoffel Tensor.
3.   3 C. The Theory of the Gravitation-Field
     1.   3.1 § 13. Equation of motion of a material point in a gravitation-field. Expression for the field-components of gravitation.
     2.   3.2 § 14. The Field-equation of Gravitation in the absence of matter.
     3.   3.3 § 15. Hamiltonian Function for the Gravitation-field. Laws of Impulse and Energy.
     4.   3.4 § 16. General formulation of the field-equation of Gravitation.
     5.   3.5 § 17. The laws of conservation in the general case.
     6.   3.6 § 18. The Impulse-energy law for matter as a consequence of the field-equations.
4.   4 D. The "Material" Phenomena.
     1.   4.1 § 19. Euler's equations for frictionless adiabatic liquid.
     2.   4.2 § 20. Maxwell's Electro-Magnetic field-equations for the vacuum.

5.   5 E. § 21. Newton's Theory as a First Approximation.
   1.   5.1 § 22. Behaviour of measuring rods and clocks in a statical gravitation-field. Curvature of light-rays. Perihelion-motion of the paths of the Planets.

General Theory of Relativity

"The Field Equations of Gravitation"
A. Einstein
November 25, 1915

The paper can be found online at:

In German
"Die Feldgleichungen der Gravitation,"
Preussische Akademie der Wissenschaften, Sitzungsberichte, Nov. 25, 1915 (part 2), 844–847, as accessed on the Internet in January 2015, go to:

https://ia601403.us.archive.org/25/items/sitzungsberichte1915deut/sitzungsberichte1915deut.pdf

In English
"The Field Equations of Gravitation" (1915)
by Albert Einstein, translated from German by Wikisource, as accessed on February 23, 2015, go to:

https://en.wikisource.org/?curid=735695

The Foundation of the Generalized Theory of Relativity (1916)

Original German
as accessed in Janurary, 2015, go to:
http://www.physik.uni-augsburg.de/annalen/history/einstein-papers/1916_49_769-822.pdf

1920 English translation
by M.N. Saha and S.N. Bose, as accessed in Jan 2015, go to:
https://archive.org/details/principleofrelat00eins

Foundation of the Generalised Theory of Relativity

By A. Einstein, 1916 [wikipedia source version]

[p. 769]

The theory which is sketched in the following pages forms the most wide-going generalization conceivable of what is at present known as "the theory of Relativity;" this latter theory I differentiate from the former "Special Relativity theory," and suppose it to be known. The generalization of the Relativity theory has been made much easier through the form given to the special Relativity theory by *Minkowski*, which mathematician was the first to recognize clearly the formal equivalence of the space like and time-like co-ordinates, and who made use of it in the building up of the theory. The mathematical apparatus useful for the general relativity theory, lay already complete in the "Absolute Differential Calculus", which were based on the researches of Gauss, Riemann and Christoffel on the non-Euclidean manifold, and which have been shaped into a system by Ricci and Levi-Civita, and already applied to the problems of theoretical physics. I have in part B of this communication developed in the simplest and clearest manner, all the supposed mathematical auxiliaries, not known to Physicists, which will be useful for our purpose, so that, a study of the mathematical literature is not necessary for an understanding of this paper. Finally in this place I thank my friend Grossmann, by whose help I was not only spared the study of the mathematical literature pertinent to this subject, but who also aided me in the researches on the field equations of gravitation. [page break]

[p. 770]

A. Principal considerations about the Postulate of Relativity.

§ 1. Remarks on the Special Relativity Theory.

The special relativity theory rests on the following postulate which also holds valid for the Galileo-Newtonian mechanics.

If a co-ordinate system K be so chosen that when referred to it the physical laws hold in their simplest forms, *these* laws would be also valid when referred to another system of co-ordinates K' which is subjected to an uniform translational motion relative to K. We call this postulate "The Special Relativity Principle." By the word special, it is signified that the principle is limited to the case, when K' has *uniform translatory motion* with reference to K, but the equivalence of K and K' does not extend to the case of *non-uniform* motion of K' relative to K.

The Special Relativity Theory does not differ from the classical mechanics through the assumption of this postulate, but only through the postulate of the constancy of light-velocity in vacuum which, when combined with the special relativity postulate, gives in a well-known way, the relativity of synchronism as well as the Lorentz transformation, with all the relations between moving rigid bodies and clocks.

The modification which the theory of space and time has undergone through the special relativity theory, is indeed a profound one, but *a* weightier point remains untouched. According to the special relativity theory, the theorems of geometry are to be looked upon as the laws about any possible relative positions of solid bodies at rest, and more generally the theorems of kinematics, as theorems which describe the relation between measurable bodies and clocks. Consider two material points of a solid body at rest; then according to these conceptions there corresponds to these points a wholly definite extent of length, independent of kind, position, orientation and time of the body.

Similarly let us consider two positions of the pointers of a clock which is at rest with reference to a co-ordinate system; then to these positions, there always corresponds a time-interval of a definite length, independent of time and place. It would be soon shown that the general relativity theory can not hold fast to this simple physical significance of space and time. [page break]

[p. 771]

§ 2. About the reasons which explain the extension of the relativity-postulate.

To the classical mechanics (no less than) to the special relativity theory, is attached an epistemological defect, which was perhaps first clearly pointed out by E. Mach. We shall illustrate it by the following example; Let two fluid bodies of equal kind and magnitude swim freely in space at such a great distance from one another (and from all other masses) that only that sort of gravitational forces are to be taken into account which the parts of *any* of these bodies exert upon each other. The distance of the bodies from one another is invariable. The relative motion of the different parts of each body is not to occur. But each mass is seen to rotate by an observer at rest relative to the other mass round the connecting line of the masses with a constant angular velocity (definite relative motion for both the masses). Now let us think that the surfaces of both the bodies $S_1(S_1$ and $S_2)$ are measured with the help of measuring rods (relatively at rest); it is then found that the surface of S_1 is a sphere and the surface of the other is an ellipsoid of rotation. We now ask, why is this difference between the two bodies? An answer to this question can only then be regarded as satisfactory[1] from the epistemological standpoint when the thing adduced as the cause is an *observable fact of experience*. The law of causality has the sense of a definite statement about the world of experience only when *observable facts* alone appear as causes and effects.

The Newtonian mechanics does not give to this question any satisfactory answer. For example, it says:— The laws of mechanics hold true for a space R_1 relative to which the body S_1 is at rest, not however for a space relative to which S_2 is at rest.

The Galiliean space, which is here introduced is however only a *purely imaginary* cause, not an observable thing. It is thus clear that the Newtonian mechanics [page break]

[p. 772]

does not, in the case treated here, actually fulfil the requirements of causality, but produces on the mind a fictitious complacency, in that it makes responsible a wholly imaginary cause R_1 for the different behaviours of the bodies S_1 and S_2 which are actually observable.

A satisfactory explanation to the question put forward above can only be thus given:— that the physical system composed of S_1 and S_2 shows for itself alone no conceivable cause to which the different behaviour of S_1 and S_2 can be attributed. The cause must thus lie *outside* the system. We are therefore led to the conception that the general laws of motion which determine specially the forms of S_1 and S_2 must be of such a kind, that the mechanical behaviour of S_1 and S_2 must be essentially conditioned by the distant masses, which we had not brought into the system considered. These distant masses, (and their relative motion as regards the bodies under consideration) are then to be looked upon as the seat of the principal observable causes for the different behaviours of the bodies under consideration. They take the place of the imaginary cause R_1. Among all the conceivable spaces R_1 and R_1 etc. moving in any manner relative to one another, there is a priori, no one set which can be regarded as affording greater advantages, against which the objection which was already raised from the standpoint of the theory of knowledge cannot be again revived. *The laws of physics must be so constituted that they should remain valid for any system of co-ordinates moving in any manner.* We thus arrive at an extension of the relativity postulate.

Besides this momentous epistemological argument, there is also a well-known physical fact which speaks in favour of an extension of the relativity theory. Let there be a Galiliean co-ordinate system K relative to which (at least in the four-dimensional region considered) a mass at a sufficient distance from other masses move uniformly in a line. Let K' be a second co-ordinate system which has a *uniformly accelerated* motion relative to K. Relative to K' any mass at a sufficiently great distance experiences an accelerated motion such that its acceleration and its direction of acceleration is independent of its material composition and its physical conditions.

Can any observer, at rest relative to K′, [page break]

[p. 773]

then conclude that he is in an actually accelerated reference-system? This is to be answered in the negative; the above-named behaviour of the freely moving masses relative to K′ can be explained in as good a manner in the following way. The reference-system K′ has no acceleration. In the space-time region considered there is a gravitation-field which generates the accelerated motion relative to K′.

This conception is feasible, because to us the experience of the existence of a field of force (namely the gravitation field) has shown that it possesses the remarkable property of imparting the same acceleration to all bodies.[2] The mechanical behaviour of the bodies relative to K′ is the same as experience would expect of them with reference to systems which we assume from habit as stationary; thus it explains why from the physical stand-point it can be assumed that the systems K and K′ can both with the same legitimacy be taken as at rest, that is, they will be equivalent as systems of reference for a description of physical phenomena.

From these discussions we see, that the working out of the general relativity theory must, at the same time, lead to a theory of gravitation; for we can "create" a gravitational field by a simple variation of the co-ordinate system. Also we see immediately that the principle of the constancy of light-velocity must be modified, for we recognise easily that the path of a ray of light with reference to K′ must be, in general, curved, when light travels with a definite and constant velocity in a straight line with reference to K.

§ 3. The time-space continuum. Requirements of the general Co-variance for the equations expressing the laws of Nature in general.

In the classical mechanics as well as in the special relativity theory, the co-ordinates of time and space have an immediate physical significance; when we say that any arbitrary point has X_1 as its X_1 co-ordinate, it signifies [page break]

[p. 774]

that the projection of the point-event on the X_1-axis ascertained by means of a solid rod according to the rules of Euclidean Geometry is reached when a definite measuring rod, the unit rod, can be carried X_1, X_2, X_3, X_4 times from the origin of co-ordinates along the X_1 axis. A point having $X_4 = t$ as the X_1 co-ordinate signifies that a unit clock which is adjusted to be at rest relative to the system of co-ordinates, and coinciding in its spatial position, with the point-event and set according to some definite standard has gone over $X_4 = t$ periods before the occurrence of the point-event.[3]

This conception of time and space is continually present in the mind of the physicist, though often in an unconscious way, as is clearly recognised from the role which this conception has played in physical measurements. This conception must also appear to the reader to be lying at the basis of the second consideration of the last paragraph and imparting a sense to these conceptions. But we wish to show that we are to abandon it and in general to replace it by more general conceptions in order to be able to work out thoroughly the postulate of general relativity,— the case of special relativity appearing as a limiting case when there is no gravitation.

We introduce in a space, which is free from Gravitation-field, a Galiliean Co-ordinate System K (x, y, z, t) and also, another system K' (y', y', z', t') rotating uniformly relative to K. The origin of both the systems as well as their Z-axes might continue to coincide. We will show that for a space-time measurement in the system K', the above established rules for the physical significance of time and space can not be maintained. On grounds of symmetry it is clear that a circle round the origin in the X-Y plane of K, can also be looked upon as a circle in the X'-Y' plane of K'. Let us now think of measuring the circumference and the diameter of these circles, with a unit measuring rod (infinitely small compared with the radius) and take the quotient of both the results of measurement. If this experiment be carried out with a measuring rod at rest relatively to the Galiliean system [page brake]

[p. 775]

K we would get π, as the quotient. The result of measurement with a rod relatively at rest as regards K' would be a number which is greater than π. This can be seen easily when we regard the whole measurement-process from the system K and remember that the rod placed on the periphery suffers a Lorentz-contraction, not however when the rod is placed along the radius. Euclidean Geometry therefore does not hold for the system K'; the above fixed conceptions of co-ordinates which assume the validity of Euclidean Geometry fail with regard to the system K'. We cannot similarly introduce in K' a time corresponding to physical requirements, which will be shown by all similarly prepared clocks at rest relative to the system K'. In order to see this we suppose that two similarly made clocks are arranged one at the centre and one at the periphery of the circle, and considered from the stationary system K. According to the well-known results of the special relativity theory it follows — (as viewed from K) — that the clock placed at the periphery will go slower than the second one which is at rest. The observer at the common origin of co-ordinates who is able to see the clock at the periphery by means of light will see the clock at the periphery going slower than the clock beside him. Since he cannot allow the velocity of light to depend explicitly upon the time in the way under consideration he will interpret his observation by saying that the clock on the periphery "actually" goes slower than the clock at the origin. He cannot therefore do otherwise than define time in such a way that the rate of going of a clock depends on its position.

We therefore arrive at this result. In the general relativity theory time and space magnitudes cannot be so defined that the difference in spatial co-ordinates can be immediately measured by the unit-measuring rod, and time-like co-ordinate difference with the aid of a normal clock.

The means hitherto at our disposal, for placing our co-ordinate system in the time-space continuum, in a definite way, therefore completely fail and [page break]

[p. 776]

it appears that there is no *other* way which will enable us to fit the co-ordinate system to the four-dimensional world in such a way, that by it we can expect to get a specially simple formulation of the laws of Nature. So that nothing remains for us but to regard all conceivable[4] co-ordinate systems as equally suitable for the description of natural phenomena. This amounts to the following law:—

That in general, Laws of Nature are expressed by means of equations which are valid for all co-ordinate systems, that is, which are covariant for all possible transformations. It is clear that a physics which satisfies this postulate will be unobjectionable from the standpoint of the general relativity postulate. Because among *all* substitutions there are, in every case, contained those, which correspond to all relative motions of the co-ordinate system (in three dimensions). This condition of general covariance which takes away the last remnants of physical objectivity from space and time, is a natural requirement, as seen from the following considerations. All our well-substantiated space-time propositions amount to the determination of space-time coincidences. If, for example, the event consisted in the motion of material points, then, for this last case, nothing else are really observable except the encounters between two or more of these material points. The results of our measurements are nothing else than well-proved theorems about such coincidences of material points, of our measuring rods with other material points, coincidences between the hands of a clock, dial-marks and point-events occurring at the same position and at the same time.

The introduction of a system of co-ordinates serves no other purpose than an easy description of totality of such coincidences. We fit to the world our space-time variables X_1, X_2, X_3, X_4 such that to any and every point-event corresponds a system of values of $X_1 \ldots X_4$. Two coincident point-events correspond to the same [page break]

[p. 777]

value of the variables $X_1 \ldots X_4$; *i.e.*, the coincidence is characterised by the equality of the co-ordinates. If we now introduce any four functions X'_1, X'_2, X'_3, X'_4 as coordinates, so that there is an unique correspondence between them, the equality of all the four co-ordinates in the new system will still be the expression of the space-time coincidence of two material points. As the purpose of all physical laws is to allow us to remember such coincidences, there is a priori no reason present, to prefer a certain co-ordinate system to another; *i.e.*, we get the condition of general covariance.

§ 4. Relation of four co-ordinates to spatial and temporal measurements. Analytical expression for the Gravitation-field.

I am not trying in this communication to deduce the general Relativity-theory as the simplest logical system possible, with a minimum of axioms. But it is my chief aim to develop the theory in such a manner that the reader perceives the psychological naturalness of the way proposed, and the fundamental assumptions appear to be most reasonable according to the light of experience. In this sense, we shall now introduce the following supposition; that for an infinitely small four-dimensional region, the relativity theory is valid in the special sense when the axes are suitably chosen.

. . .

About the Author

Walter Dolen is an author who uses the scientific method[171] to research the material for his books. He has researched and written on science, chronology, philosophy, psychology, theology, religion, sex differences, feminism and so forth. The author questions everything and from this he writes his books. For more info on the author see:

www.walterdolen.com or www.walterdolen.ws

---

[171] (1) Perceive a problem; (2) examine and analyze all the available evidence; (3) examine and imagine different hypotheses in attempt to solve the problem in a logical manner; (4) form a theory that answers the problem; (5) test the theory; (6) always have an open mind for better theories or answers to the problem; (7) change the theory if new evidence is inconsistent to your prior theory.